A Primer of Conservation Biology

A Primer of Conservation Biology
Second Edition

Richard B. Primack

BOSTON UNIVERSITY

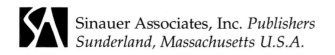

Sinauer Associates, Inc. *Publishers*
Sunderland, Massachusetts U.S.A.

The Cover

The hyacinth macaw (*Anodorhynchus hyacinthinus*) is an endangered species of South America, with only 3000 to 7000 birds remaining in the wild. Conservation efforts include protecting known breeding sites, developing ecotourist facilities at such sites to provide local revenue, and stopping the illegal trade in macaws. Photograph copyright © Frans Lanting/Minden Pictures.

Library of Congress Cataloging-in-Publication Data

Primack, Richard B., 1950-
    A primer of conservation biology / Richard B. Primack.—2nd ed.
        p. cm.
    Includes bibliographical references (p.).
    ISBN 0-87893-732-3 (paper)
    1. Conservation biology.   I. Title.

QH75 .P7525   2000
333.95'16—dc21   00-029693

Printed in U.S.A.
6 5 4 3 2 1

This book is dedicated to those who teach conservation biology, ecology, and environmental sciences, whose efforts will inspire future generations to find the right balance between protecting biological diversity and providing for human needs.

Contents

Preface

*T*he image of biological diversity as a treasure is symbolically highlighted in a recent finding that certain plants have the ability to concentrate gold when grown on the surface of old mines. Conservation biology seeks to protect the Earth's most precious treasure— its biological diversity. The field emerged during the last 15 years as a major new discipline to address the alarming loss of biological diversity throughout the world. Though still in its formative stages, conservation biology is evolving at a rapid pace and it is attracting increasing scientific, governmental, and popular attention.

Evidence of the explosive increase of interest in this new scientific field is shown by the rapidly increasing membership in the Society for Conservation Biology, the great intellectual excitement displayed in many journals and newsletters, and the large numbers of new edited books and advanced texts on the subject that appear almost weekly. In the United States, a major grant by one of the founders of the Intel Corporation is being used to establish a Center for Applied Biodiversity Science, and a Global Biodiversity Information Facility is being developed to coordinate the efforts of 29 major industrialized countries.

Such interest continues to extend to university students, who have enrolled enthusiastically and in large numbers in introductory conservation biology courses. The publication of the first edition of the *Primer* sought to fill the need for a "quick" guide for those who wanted a basic familiarity with conservation biology. This second edition of the *Primer* attempts to meet that same goal—to provide a brief but thorough introduction to the major concepts and problems of the field. Like its predecessor, it is designed for use in short courses in conservation biology, and can also be used as a supplemental text for general biology, ecology, wildlife biology, and environmental policy courses. It is also intended to serve as a concise guide for professionals who require a background in the subject, but have no need for in-depth case studies and discussions.

In keeping with the international approach of conservation biology, this edition of the *Primer* continues my attempt, begun with my first text, *Essentials of Conservation Biology* (1993; second edition 1998), to make the field accessible to as wide an audience as possible. With the assistance of Marie Scavotto and other members of the staff of Sinauer Associates, I have arranged an active translation program. Although the *Essentials* text was published in German as *Naturschutzbiologie* in 1995, and was followed by a translation into Chinese, it became clear to me that the best way to make the material more widely available was to create regional or country-specific translations, identifying scientists in each location to become coauthors, and to add case studies, examples, and illustrations from their own countries and regions that would be more relevant to the intended audience. To that end, in the past four years, editions of the *Primer* have appeared in Japanese (with Hiromi Kobori), in Indonesian (with Jatna Supriatna and others), in Vietnamese (with Pham Binh Quyen, Vo Quy, and Hoang Van Thang), and in Korean (with Dowon Lee and others). Editions of the *Essentials* and *Primer* texts in French, Greek, Hungarian, Czech, Arabic, Spanish, Italian, Brazilian Portuguese, and for India, are planned for the near future. The MacArthur Foundation and the Conservation, Food and Health Foundation have provided funds for several of these translations. It is my desire that these translations will help conservation biology develop as a discipline with a global scope.

I hope that readers of this book will want to find out more about the current extinction crisis facing species and biological communities and how they can take action to halt it. I encourage readers to take conservation biology's activist spirit to heart—use the Appendix to find organizations that provide information and suggestions for how each of us can help. If readers gain a greater appreciation for the goals, methods, and importance of conservation biology, and if they are moved to make a difference in their everyday lives to help preserve our biological diversity, this primer will have served its purpose.

Acknowledgments

I sincerely appreciate the contributions of Elizabeth Platt, Phil Cafaro, April Algaier Stern, Linus Chen, and others acknowledged in *A Primer of Conservation Biology*, First Edition, and *Essentials of Conservation Biology*, Second Edition, from which much of the material in this text was drawn. Their assistance was invaluable, and this primer would not have been possible without it. Portions of this text were reviewed by Sevima Aktay, Phil Cafaro, Daniel Janzen, Tigga Kingston, Charles Munn, Bruce Rich, and Lisa Sorenson; numerous other individuals also provided specific information in their areas of expertise. Linus Chen assisted in the editing and provided much valuable commentary on draft chapters. Lisa Delissio kept the computer files in good order. Special thanks are due to Kamaljit Bawa and Margaret Primack for their constant encouragement. I would also like to thank Boston University and Harvard University for providing me, while on sabbatical leave, with the excellent facilites and environment that made this project possible. And finally, thanks to Andy Sinauer, Kerry Falvey, Chris Small, David McIntyre, and the other members of the Sinauer Associates staff, who performed their usual magic of transforming my piles of notes, email messages, and hand-drawn sketches into an elegant and useful book.

RICHARD PRIMACK
BOSTON UNIVERSITY
APRIL 2000

Chapter *1*

Conservation Biology and Biological Diversity

*A*round the globe, biological communities that took millions of years to develop are being devastated by human actions. The list of transformations of natural systems that are directly related to human activities is long. Vast numbers of species have declined rapidly, some to the point of extinction, as a result of excessive hunting, habitat destruction, and an on- slaught of introduced predators and competitors (Heywood 1995; Lawton and May 1995). Natural hydrologic and chemi- cal cycles have been disrupted by the clearing of land, which causes billions of tons of topsoil to erode and wash into rivers, lakes, and oceans each year. Genetic diversity has decreased, even among species with otherwise healthy populations. The very climate of our planet Earth has been disrupted by a com- bination of atmospheric pollution and deforestation. The pres- ent threats to biological diversity are unprecedented: Never before in the history of life have so many species been threat- ened with extinction in so short a period of time. These threats to biological diversity are accelerating due to the demands of a rapidly increasing human population and its rising material consumption. This dire situation is exacerbated by the unequal distribution of the world's wealth and the crushing poverty of many of the tropical countries that have an abundance of species. Moreover, many of the threats to biological diversity are synergistic; that is, several independent factors, such as acid rain, logging, and over-hunting, combine exponentially to make the situation even worse (Myers 1987). What is bad for biological diversity will almost certainly be bad for human populations because humans are dependent on the natural environment for air and water, raw materials, food, medicines, and other goods and services.

While some people may feel discouraged by the avalanche of species extinctions and habitat degradation occurring in the world today, it is also possible to feel challenged by the need to do something to stop the destruction. As Peter Raven, director of the Missouri Botanical Garden, has said regarding the loss of biological diversity: "You can think about it on a worldwide basis, and then it becomes discouraging and insoluble, or you can think about it in terms of specific opportunities, seize those opportunities, and reduce the problem to a more manageable size" (quoted in Tangley 1986). The next few decades will determine how much of the world's biological diversity will survive. The efforts now being made to save species, establish new conservation areas, and protect existing national parks will determine which of the world's species and biological communities are preserved for the future. **Conservation biology** is the scientific discipline that has developed out of these efforts. It brings together people and knowledge from many different fields to address the bio-diversity crisis. In the future, people may look back on the closing years of the twentieth century and the early years of the twenty-first as a time when a relative handful of determined people saved numerous species and biological communities from extinction.

Conservation Biology's Interdisciplinary Approach: A Case Study

Macaws are familiar to most people as large, brightly colored parrots with a mischievous, amiable intelligence that makes them the clowns of the bird world. Yet for all their popularity as pets and performers, macaws are in trouble in the wild. The efforts to protect macaw species serve as a powerful illustration of the interdisciplinary nature of conservation biology.

Nine of the sixteen species of macaws that reside in the tropical forests of South America are endangered, with at least one species, the Spix's macaw, in imminent danger of extinction. Throughout their range, all species have declined dramatically in numbers as a result of a combination of factors: the systematic collection of wild birds for pets, hunting by local people, and forest destruction.

Given the variety of factors contributing to the macaws' decline, it is perhaps not surprising that a multifaceted approach is required for the conservation of these birds.

In spite of their popularity as pets, very little was known about macaws in the wild until recently, in part because they are abundant only in remote regions of tropical forests. One such area is the rain forest of southeastern Peru, where wildlife biologists from the Wildlife Conservation Society (WCS) have been conducting research since 1984 on eight species of macaws (Munn 1992, 1994; Diamond 1999). At first, these scientists concentrated on learning the species' basic biology: their dietary requirements, reproductive habits, and other basic needs. In the course of the study, however, a previously reported, unique aspect of macaw behavior quickly became evident: Large numbers of macaws congregate at isolated clay cliff faces to eat clay (Figure 1.1). An initial hypothesis of this puzzling behavior was that macaws needed the trace minerals in the clay to supplement their diet of seeds and fruits. However, a biochemical investigation demonstrated that the clay is actually needed to detoxify poisonous chemicals contained in the seeds the birds eat. Despite their small area in the overall landscape, the clay licks represent a vital, possibly irreplaceable resource for the macaws that visit them.

Even before this discovery, biologists had been aware that urgent action was needed to halt the decline of the macaws. Mining and timber companies are encroaching on the macaws' habitat. Hunting

(A)

(B)

1.1 (A) Red-and-green macaws (*Ara chloroptera*) gather at cliffs to feed on exposed clay. (B) A blue-and-yellow macaw (*Ara ararauna*) uses a tree cavity as a nest site. (Photographs © Charles Munn.)

of macaws for food and collecting for the pet trade by local Indians and more recent settlers are also contributing to the decline in the species' populations. Though political lobbying by wildlife advocacy organizations is leading toward a ban on international trade in macaws by most tropical American countries and imports of macaws by the United States, the combination of markets within the tropical countries themselves, loopholes in the laws, and a growing black market continue to put pressure on wild macaw populations. Active efforts to protect macaw habitat and discourage collecting and hunting are still needed. The discovery of the macaws' dependence on clay licks emphasizes the fact that any protected areas established for the birds must include key resources to assure the macaws' continued survival. Moreover, the macaws' tendency to congregate in specific locations to acquire these resources makes them especially vulnerable to hunters and collectors. Conservation biologists thus are seeking ways to discourage the collection of wild birds for pets and the hunting of birds for meat.

As a result of the research by the WCS biologists and other scientists working in the area, several steps are being taken to protect macaws and their habitat, including the clay licks. New and proposed national parks in Peru, such as Manu National Park and the 1.5 million hectare* Tambopata–Candamo Reserved Zone, deliberately include many of the macaw clay licks and surrounding forest habitat. Management of the parks is being designed to permit a degree of sustainable development in order to provide employment for local people and to support the parks economically.

Key elements in this strategy are sustainable harvesting of Brazil nuts, establishment of commercial zones at the parks' peripheries for small-scale streambed gold mining, and ecotourism centered on the macaw clay licks. With assistance from the wildlife biologists, tourist lodges have been built in the parks and on the park borders, many of which are owned, operated, and staffed by local people. Local people have been trained as field guides, research assistants, and park employees. Information from the ongoing research program is being incorporated into nature tours, educational videos, and tourist brochures. Some local people already see that the macaws are the key to their economic future rather than just their next meal, and as a result, they are taking an active role in maintaining the beauty and environmental quality of the parks. Wildlife biologists working in the area are also helping these people gain legal title to their traditional lands so they are able to take part in shaping the long-term development of the region and benefit from the

---

*For an explanation of the term *hectare* and other measurements, see Table 1.1.

TABLE 1.1   Some useful units of measurement

Length	
1 meter (m)	1 m = 39.4 inches
1 kilometer (km)	1 km = 1000 m = 0.62 miles
1 centimeter (cm)	1 cm = 1/100 m = 0.39 inches
1 millimeter (mm)	1 mm = 1/1000 m = 0.039 inches
Area	
square meter (m^2)	Area encompassed by a square, each side of which is 1 meter
1 hectare (ha)	1 ha = 10,000 m^2 = 2.47 acres; 100 ha = 1 square kilometer (km^2)
Mass	
1 kilogram (kg)	1 kg = 2.2 pounds
1 gram (g)	1 g = 1/1000 kg = 0.035 ounce
Temperature	
degree Celsius (°C)	0°C = 32° Fahrenheit (°F) (freezing point of water) 100°C = 212°F (boiling point of water) 23°C = 72°F ("room temperature")

highly profitable ecotourism industry. Although the Tambopata–Candamo Reserved Zone programs, which involve local people in park development, seem promising, they are not without controversy: Other biologists working in the region have proposed excluding all people from the parks in order to maintain them as wilderness areas with minimal human impact.

The establishment of protected areas for macaws has been a boon to conservation activities and may ultimately be instrumental in restoring these birds. Two recent discoveries about macaw reproductive behavior may also prove useful in rebuilding macaw populations. First, ornithologists working elsewhere in South America have discovered that macaws have exacting requirements for nest sites, which are usually cavities in the trunks of large trees. Suitable cavities are apparently few and far between, and if none are available, pairs will not breed. To overcome the scarcity of nesting sites, WCS scientists designed wooden and plastic nest boxes that can be attached to tree trunks. Three species, the scarlet, blue-and-gold, and hyacinth macaws, have shown a willingness to accept these boxes as nests. Distribution of the nest boxes throughout the parks may not only help to increase the population density of the species but may also allow macaws to breed in logged forests, from which large trees have been removed. Second, researchers have determined that although macaws often lay two eggs per nest, only the older nestling typically survives. The researchers found that they can remove the younger nestling and raise it by hand. Once these captive-fostered juveniles fledge, they join the wild flock at the clay licks.

These and other techniques that have grown out of the ongoing research are being incorporated into conservation efforts in Bolivia and Brazil, where macaw species are highly endangered. The most significant outcome of the project for conservation biology as a whole, however, lies in the interdisciplinary element at work in this endeavor. Active, dedicated scientists with a long-term commitment to investigating and protecting an important but poorly known group of species recognized the need to become politically active in order to establish legally protected areas, control international trade in the species, and develop sustainable economic opportunities for local people in the growing nature-tourism industry. By incorporating their research results into a flexible management plan, in this case involving artificial nest construction and captive rearing of nestlings, the researchers were able to make progress toward their goal. Furthermore, by publicizing their results in highly visible publications such as *National Geographic* and in videos produced for international television, the researchers made it possible for the techniques developed by this project to be transferred to conservation efforts in other countries, enhancing the value of their outstanding work. The lessons for conservation biology are clear: By attacking a problem from several different angles, researchers can address the underlying biological, economic, sociological, and management problems that threaten species.

Why Is Conservation Biology Needed?

Conservation biology is a multidisciplinary science that has developed in response to the crisis confronting biological diversity today (Wilson 1992; Meffe and Carroll 1997; Primack 1998). Conservation biology has three goals: first, to investigate and describe the diversity of the living world; second, to understand the effects of human activities on species, communities, and ecosystems; and third, to develop practical interdisciplinary approaches to protecting and restoring biological diversity.

Conservation biology arose because none of the traditional applied disciplines are comprehensive enough by themselves to address the critical threats to biological diversity. Agriculture, forestry, wildlife management, and fisheries biology have been concerned primarily with developing methods for managing a small range of species for the marketplace and for recreation. Although these disciplines are increasingly considering conservation issues, they generally have not addressed the need for protecting the full range of species found in biological communities, or have regarded it as a secondary issue. Conservation biology complements these applied disciplines by providing a more general theoretical approach to the protection of biological diversity; it differs from those disciplines in having the long-term preservation of entire biological communities as its primary consideration, with economic factors often secondary.

The academic disciplines of population biology, taxonomy, ecology, landscape ecology, and genetics constitute the core of conservation biology, and many conservation biologists have been drawn from their ranks. In addition, many leaders in conservation biology have come from zoos and botanical gardens, bringing with them experience in maintaining and propagating species in captivity. Because much of the biodiversity crisis arises from human pressures, conservation biology also incorporates ideas and expertise from a broad range of fields outside of biology (Figure 1.2). For example, environmental law and policy provides the basis for governmental protection of rare and endangered species and critical habitats. Environmental ethics provides a rationale for preserving

1.2 Conservation biology represents a synthesis of many basic sciences (left) that provide principles and new approaches for the applied fields of resource management (right). The experiences gained in the field in turn influence the direction of the basic sciences. (After Temple 1991.)

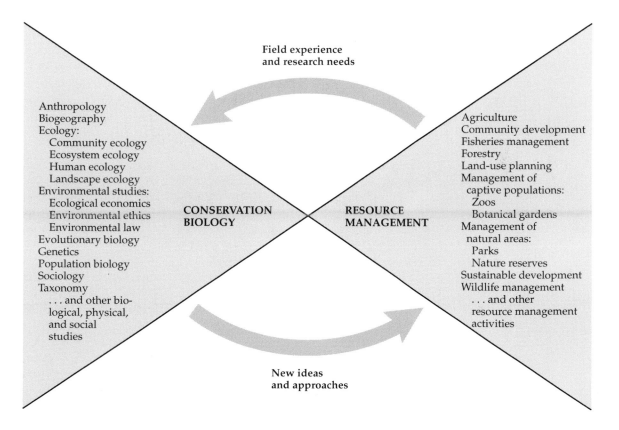

Field experience and research needs

Anthropology
Biogeography
Ecology:
    Community ecology
    Ecosystem ecology
    Human ecology
    Landscape ecology
Environmental studies:
    Ecological economics
    Environmental ethics
    Environmental law
Evolutionary biology
Genetics
Population biology
Sociology
Taxonomy
    . . . and other bio-
    logical, physical,
    and social
    studies

CONSERVATION BIOLOGY

RESOURCE MANAGEMENT

Agriculture
Community development
Fisheries management
Forestry
Land-use planning
Management of
    captive populations:
    Zoos
    Botanical gardens
Management of
    natural areas:
    Parks
    Nature reserves
Sustainable development
Wildlife management
    . . . and other
    resource management
    activities

New ideas and approaches

species and habitats. Social sciences such as anthropology, sociology, and geography provide insight into how people can be encouraged and educated to protect the natural resources and species found in their immediate environment. Ecological economists provide analyses of the economic value of biological diversity to support arguments for preservation. Ecosystem ecologists and climatologists monitor the biological and physical characteristics of the environment and develop models to predict environmental responses to disturbance.

In many ways, conservation biology is a crisis discipline. Decisions on conservation issues are being made every day, often with limited information and under severe time pressure. Conservation biology tries to provide answers to specific questions applicable to actual situations. Such questions are raised in the process of determining the best strategies for protecting rare species, designing nature reserves, creating management plans for parks and multiple-use areas, and reconciling conservation concerns with the needs of local people and governments. Conservation biologists and scientists in related fields are well suited to provide the advice that governments, businesses, and the general public need to make crucial decisions. Although some conservation biologists would prefer not to make recommendations without detailed knowledge of the specifics of a case, the urgency of many situations necessitates informed decisions based on certain fundamental principles of biology. This book describes those principles and gives examples of how they may be factored into policy making.

The Philosophical Background of Conservation Biology

The need for the conservation of biological diversity has been recognized for decades, even centuries, in North America, Europe, and other regions of the world. Religious and philosophical beliefs concerning the value of protecting species and wilderness are found in many cultures worldwide (Hargrove 1989; Callicott 1994). Many religions emphasize the need for people to live in harmony with nature and to protect the living world because it is a divine creation. In the United States, philosophers such as Ralph Waldo Emerson and Henry David Thoreau saw wild nature as an important element in human moral and spiritual development (Callicott 1990). Wilderness advocates such as John Muir and Aldo Leopold argued for preserving natural landscapes and maintaining the health of natural ecosystems. A similar perspective is put forward by the **Gaia hypothesis**, which views the Earth as having the properties of a "superorganism" whose biological, physical, and chemical components interact to regulate characteristics of the atmosphere and climate (Lovelock 1988). Modern proponents of wilderness, such as environmental activists and members of the deep ecology movement, which is discussed later in this chapter, often advocate reduction or complete cessation of practices and industries that disrupt the normal interaction of the Earth's components.

Paralleling these preservationist and ecological orientations, an influential forester, Gifford Pinchot (1865–1946), developed the idea that commodities and qualities found in nature, including timber, fodder, clean water, wildlife, species diversity, and even beautiful landscapes, can be considered **natural resources** and that the goal of management is to use these natural resources for the greatest good of the greatest number of people for the longest time. Pinchot's and Leopold's ideas have been combined and extended by the concept of **ecosystem management**, which places the highest management priority on the health of ecosystems and wild species (Noss and Cooperrider 1994). The current paradigm of **sustainable development** also advocates an approach similar to Pinchot's: developing natural resources to meet present human needs in a way that does not harm biological communities and considers the needs of future generations as well (Lubchenco et al. 1991).

The modern discipline of conservation biology rests on several underlying ethical beliefs that are generally agreed upon by members of the discipline (Soulé 1985). These ethical beliefs suggest research approaches and practical applications. Although not all of these statements are accepted unequivocally by all conservation biologists, the acceptance of one or two is sufficient rationale for people to become involved in conservation efforts.

1. *The diversity of species and biological communities should be preserved.* In general, people enjoy biological diversity. The hundreds of millions of visitors each year to zoos, national parks, botanical gardens, and aquariums are testimony to the general public's interest in observing different species and biological communities. Genetic variation within species also has popular appeal, as shown by dog shows, cat shows, agricultural expositions, and flower exhibitions. It has even been speculated that humans have a genetic predisposition to like biological diversity, called **biophilia** (Wilson 1984; Kellert and Wilson 1993). Biophilia would have been advantageous for the hunting-and-gathering lifestyle that humans led for hundreds of thousands of years before the invention of agriculture. High biological diversity would have provided them with a variety of foods and other resources, buffering them against environmental catastrophes and starvation.

2. *The untimely extinction of populations and species should be prevented.* The extinction of species and populations as a result of natural processes is a natural event. Through the millennia of

geological time, extinctions of species have generally been balanced by the evolution of new species. Likewise, the local loss of a population is usually offset by the establishment of a new population through dispersal. However, human activity has increased the rate of extinction a thousandfold (Lawton and May 1995). In the twentieth century, virtually all of the hundreds of known extinctions of vertebrate species, as well as the presumed thousands of extinctions of invertebrate species, have been caused by humans.

3. *Ecological complexity should be maintained.* Many of the most interesting properties of biological diversity are expressed only in natural environments. For example, complex coevolutionary and ecological relationships exist among tropical flowers, hummingbirds, and mites that live in the flowers. The mites use the hummingbirds' beaks as "buses" to go from flower to flower (Colwell 1986). Such relationships would never be suspected if the animals and plants were housed in isolation in zoos and botanical gardens. Similarly, the fascinating behaviors used by desert animals to obtain water would not be apparent if the animals were living in cages and supplied with water to drink at will. While it might be possible to preserve at least some of the diversity of flowering plants and vertebrate species in zoos and gardens, the ecological complexity that exists in natural communities would be largely lost. This provides an argument for protecting viable examples of all biological communities.

4. *Evolution should continue.* Evolutionary adaptation is the process that eventually leads to new species and increased biological diversity. Therefore, allowing populations to evolve in nature is beneficial. Human activities that limit the ability of populations to evolve, such as severely reducing a species' population size through overharvesting and eliminating unique populations, are destructive.

5. *Biological diversity has intrinsic value.* Species and the biological communities in which they live, have a value all their own, regardless of their value to human society. This value is conferred by their evolutionary history and unique ecological roles, and also by their very existence. Thus all species should be preserved.

What Is Biological Diversity?

The protection of biological diversity is central to conservation biology. The definition of biological diversity given by the World Wildlife Fund (1989) is, "the wealth of life on earth, the millions of plants, animals, and microorganisms, the genes they contain, and the intricate ecosystems they help build into the living environment." Thus biological diversity needs to be considered at three levels. Biological diversity at the species level includes the full range of organisms on Earth, from bacteria and protists through the multicellular kingdoms of plants, animals, and fungi. On a finer scale, biological diversity includes the genetic variation within species, both among geographically separated populations and among individuals within single populations. Biological diversity also includes variation in the biological communities in which species live, the ecosystems in which communities exist, and the interactions among these levels (Figure 1.3).

1.3 Biological diversity includes genetic diversity (the genetic variation found within each species), species diversity (the range of species in a given ecosystem), and community/ecosystem diversity (the variety of habitat types and ecosystem processes extending over a given region). (After Temple 1991; drawings by Tamara Sayre.)

Genetic diversity in a rabbit population

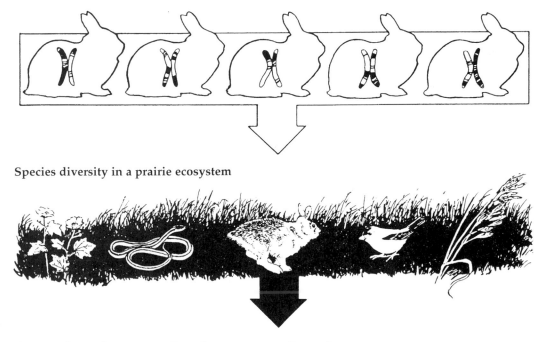

Species diversity in a prairie ecosystem

Community and ecosystem diversity across an entire region

All levels of biological diversity are necessary for the continued survival of species and natural communities, and all are important to people. Species diversity represents the range of evolutionary and ecological adaptations of species to particular environments. The diversity of species provides people with resources and resource alternatives; for example, a tropical rain forest with many species produces a wide variety of plant and animal products that can be used for food, shelter, and medicine. Genetic diversity is needed by any species in order to maintain reproductive vitality, resistance to disease, and the ability to adapt to changing conditions. Genetic diversity in domestic plants and animals is of particular value to people who work with the breeding programs that are necessary to sustain and improve modern agricultural species. Community-level diversity represents the collective response of species to different environmental conditions. Biological communities found in deserts, grasslands, wetlands, and forests support the continuity of proper ecosystem functioning, providing beneficial services such as flood control, protection from soil erosion, and filtering of air and water.

Species diversity

At each level of biological diversity—species, genetic, and community—conservation biologists study the mechanisms that alter or maintain diversity. Species diversity includes the entire range of species found on Earth. A species is generally defined in one of two ways. First, a species can be defined as a group of individuals that is morphologically, physiologically, or biochemically distinct from other groups in some characteristic (the **morphological definition of species**). Increasingly, differences in DNA sequences and other molecular markers are being used to distinguish species that look almost identical, such as bacteria. Second, a species can be distinguished as a group of individuals that can potentially breed among themselves and do not breed with individuals of other groups (the **biological definition of species**).

The morphological definition of species is the one most commonly used by **taxonomists**, biologists who specialize in the identification of unknown specimens and the classification of species. The biological definition of species is the one most commonly used by **evolutionary biologists** because it is based on measurable genetic relationships rather than on somewhat subjective physical features. In practice, however, the biological definition of species is hard to use because it requires knowledge of which individuals are actually capable of breeding with one another—information that is rarely available. As a result, practicing field biologists learn to tell species apart by how they look, sometimes calling them "**morpho-species**" or other such terms until taxonomists can give them official, Latin names (Box 1.1).

Box 1.1 **Naming and Classifying Species**

Modern taxonomy, the science of classifying living things, is creating a classification system that reflects the evolutionary relationships of species. By identifying these relationships, taxonomists help conservation biologists identify species or groups that may be evolutionarily unique or particularly worthy of conservation efforts. In modern classification,

Similar **species** are grouped into a **genus**: the Blackburnian warbler (*Dendroica fusca*) and many similar warbler species belong to the genus *Dendroica*.

Similar **genera** are grouped into a **family**: all wood warbler genera belong to the family Parulidae.

Similar **families** are grouped into an **order**: all songbird families belong to the order Passeriformes.

Similar **orders** are grouped into a **class**: all bird orders belong to the class Aves.

Similar **classes** are grouped into a **phylum**: all vertebrate classes belong to the phylum Chordata.

Similar **phyla** are grouped into a **kingdom**: all animal classes belong to the kingdom Animalia.

Modern biologists now recognize three domains with six kingdoms in the living world: the first domain and kingdom is the Bacteria and includes

Kingdom: Animalia

>1,000,000 species

Phylum: Chordata

±40,000 species

Class: Aves (birds)

8,600 species

Order: Passeriformes
    (songbirds)
5,160 species

Family: Parulidae
    (wood warblers)
125 species

Genus: *Dendroica*

28 species

Species: *Dendroica fusca*

Blackburnian warbler

Less specific

More specific

The Blackburnian warbler (*Dendroica fusca*) can be grouped with more and more other animals at successively higher levels of taxonomic organization.

Box 1.1 *(continued)*

single-celled species without a nucleus; the second domain and kingdom is the Archaea, consisting of evolutionarily distinct, bacteria-like species that often live in extreme environments; the third domain is the Eukarya, species with a nucleus, which includes the four remaining kingdoms of single-celled protists, animals, plants, and fungi. The animal kingdom has the most species, and the bacteria are the least well-known taxonomically.

Biologists throughout the world have agreed to use a standard set of names, often called scientific names or Latin names, when discussing species. This naming system, known as **binomial nomenclature**, was developed in the eighteenth century by the Swedish biologist Carolus Linnaeus. The use of scientific names avoids the confusion that can occur when the common names found in everyday language are used. The scientific name is standard across countries and languages, removing the possibility of misidentifying a species that is known by multiple names in several languages. Scientific species names consist of two words. In the scientific name for the Black-

burnian warbler, *Dendroica fusca*, *Dendroica* is the genus name and *fusca* is the specific name. The genus name is somewhat like a person's family name in that many closely related people can have the same family name (Sullivan), while the specific name is like a person's given name (Carol) within her or his family.

Scientific names are written in a standard way. The first letter of the genus name is always capitalized, whereas the species name is almost always lowercased. Scientific names are either italicized or underlined. Sometimes scientific names are followed by a scientist's name, as in *Homo sapiens Linnaeus*, indicating that Linnaeus was the person who first proposed the scientific name given to the human species. When many species in a single genus are being discussed, or if the identity of a species within a genus is uncertain, the abbreviations spp. or sp., respectively, are sometimes used (e.g., *Dendroica* spp.). If a species has no close relatives, such as the Giant Panda, it may be the only species in its own genus. Similarly, a genus that is unrelated to any other genera may form its own family.

Problems in distinguishing and identifying species are more common than many people realize (Brownlow 1996; Soltis and Gitzendanner 1999). For example, a single species may have several varieties that have observable morphological differences, yet the varieties may be similar enough that they are still considered members of a single biological species. Different breeds of dogs, such as German shepherds, collies, and beagles, all belong to one species and readily interbreed despite the conspicuous differences among them. In contrast, there are closely related "sibling species" that are very similar in morphology or physiology, yet are still biologically separate and do not interbreed. In practice, biologists often have difficulty distinguishing variation within a single species from variation between closely related species. Further complicating matters, what are otherwise distinct species may occasionally mate and produce **hybrids**, intermediate offspring that blur the distinction between species. Hybridization is particularly common among plant species in disturbed habitats. Finally, for many groups of species, the taxonomic studies needed to determine species and identify specimens have not yet been done.

The inability to clearly distinguish one species from another, whether due to similarities of characteristics or to confusion over the correct scientific name, often slows down efforts at species protection. It is difficult to write precise, effective laws to protect a species if it is not certain what name should be used. A lot more work is needed to catalogue and classify the world's species. Taxonomists have described only 10%–30% of the world's species, and many species are going extinct before they can be described. The key to solving this problem is to train more taxonomists, particularly for work in the species-rich Tropics (Raven and Wilson 1992).

Genetic diversity

Genetic diversity within a species is often affected by the reproductive behavior of individuals within populations. A **population** is a group of individuals that mate with one another and produce offspring; a species may include one or more separate populations. A population may consist of only a few individuals or millions of individuals.

Individuals within a population usually are genetically different from one another. Genetic variation arises because individuals have slightly different genes, the units of the chromosomes that code for specific proteins. The different forms of a gene are known as **alleles**, and the differences arise through **mutations**—changes that occur in the deoxyribonucleic acid (DNA) that constitutes an individual's chromosomes. The various alleles of a gene may affect the development and physiology of the individual organism differently. Crop breeders and animal breeders take advantage of this genetic variation to breed higher-yielding, pest-resistant strains of domesticated species such as wheat, corn, cattle, and poultry (Figure 1.4).

Genetic variation increases when offspring receive unique combinations of genes and chromosomes from their parents via the **recombination** of genes that occurs during sexual reproduction. Genes are exchanged between chromosomes during meiosis, and new combinations are created when chromosomes from two parents combine to form a genetically unique offspring. Although mutations provide the basic material for genetic variation, the ability of sexually reproducing species to randomly rearrange alleles in different combinations dramatically increases their potential for genetic variation.

The total array of genes and alleles in a population is referred to as its **gene pool**, while the particular combination of alleles that any individual possesses is referred to as its **genotype**. The **phenotype** of an individual represents the morphological, physiological, and biochemical characteristics that result from the expression of its genotype in a particular environment (Figure 1.5). Some characteristics of humans, such as the amount of body fat and tooth decay,

(A)

(B)

(C)

1.4 Artificial selection can produce domestic animals "tailored" to best meet the needs of humans. (A) This Australian beef cow was bred to produce abundant meat. (Photograph © John N. A. Lott/Biological Photo Service.) (B) This cow in Gimmewald, Switzerland, grazes on lush alpine grass and produces quantities of rich milk. (C) The hardy Ankole cows of Kenya can survive long droughts. Their milk, blood, and dung are all used by their herders. (Photographs in B and C © Robert E. Ford/TERRAPHOTOGRAPHICS.)

are strikingly influenced by the environment, while other characteristics, such as eye color and blood type, are determined predominantly by an individual's genotype.

The amount of genetic variability in a population is determined both by the number of genes in its gene pool that have more than one allele (referred to as **polymorphic genes**) and by the number of alleles for each polymorphic gene. The existence of a polymorphic gene allows individuals in the population to be **heterozygous** for the gene, that is, to receive a different allele of the gene from each parent. Genetic variability allows species to adapt to a changing environment, which may include, for example, higher temperatures or the outbreak of a new disease. In general, it has been found that rare species have less genetic variation than widespread species and consequently are more vulnerable to extinction when environmental conditions change.

Community and ecosystem diversity

A **biological community** is defined as the species that occupy a particular locality and the interactions among those species. Examples

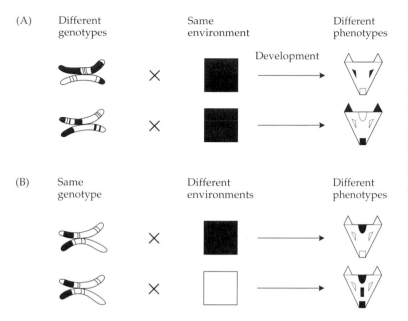

1.5 The physical, physiological, and biochemical characteristics of an individual—its phenotype—are determined by its genotype and by its environment. (A) Genetically different individuals may have different phenotypes even if they develop in the same environment. (B) Genetically similar individuals may have different phenotypes if they develop in different environments (e.g., hot vs. cold climate; abundant vs. scarce food). (After Alcock 1993.)

of communities include coniferous forests, tall-grass prairies, tropical rain forests, coral reefs, and deserts. A biological community together with its associated physical environment is termed an **ecosystem**. Within a terrestrial ecosystem, water evaporates from biological communities and the Earth's surface to fall again as rain or snow and replenish terrestrial and aquatic environments. Photosynthetic plants absorb light energy, which is used in the plants' growth. This energy is captured by animals that eat the plants, or is released as heat, both during the organisms' life cycles and after they die and decompose. Plants absorb carbon dioxide and release oxygen during photosynthesis, while animals and fungi absorb oxygen and release carbon dioxide during respiration. Mineral nutrients, such as nitrogen and phosphorus, cycle between the living and the nonliving compartments of the ecosystem.

The physical environment, especially annual cycles of temperature and precipitation, affects the structure and characteristics of a biological community, determining whether a particular site will be a forest, a grassland, a desert, or a wetland. The biological community can also alter the physical characteristics of an ecosystem. In a terrestrial ecosystem, for example, wind speed, humidity, temperature, and soil characteristics at a given location can all be affected by the plants and animals present there. In aquatic ecosystems, such physical characteristics as water turbulence and clarity, water chemistry, and water depth affect the characteristics of the associated biota, but communities such as kelp forests and coral reefs can, in turn, affect the physical environment.

Within a biological community, each species utilizes a unique set of resources that constitutes its **niche**. The niche for a plant species might consist of the type of soil on which it grows, the amount of sunlight and moisture it requires, the type of pollination system it has, and its mechanism of seed dispersal. The niche for an animal might include the type of habitat it occupies, its thermal tolerances, its dietary requirements, its home range or territory, and its water requirements. Any component of the niche may become a **limiting resource** when it restricts population size. For example, populations of bat species that have specialized roosting requirements, forming colonies only in limestone caves, may be restricted by the number of caves with the proper conditions for roosting sites.

The niche often includes the stage of succession that the species occupies. **Succession** is the gradual process of change in species composition, community structure, and physical characteristics that occurs following natural or human-caused disturbance to a biological community. Certain species are often associated with particular successional stages. For example, sun-loving butterflies and annual plants most commonly are found early in succession, in the months immediately after a gap opens in an old-growth forest. Other species, including shade-tolerant wildflowers and birds that nest in holes in dead trees, are found in late-successional stages, among mature trees in an old-growth forest. Human management patterns often upset the natural pattern of succession; grasslands that are overgrazed by cattle and forests from which all the large trees have been cut for timber typically no longer have their rare late-successional species.

The composition of communities is often affected by competition and predation. Predators often dramatically reduce the numbers of their prey species and may eliminate some species from certain habitats. Predators often indirectly increase biological diversity in a community by keeping the densities of some prey species so low that competition for resources does not occur. The number of individuals of a particular species that the resources of an environment can support is termed the **carrying capacity**. A population's numbers are often well below the carrying capacity when the species is held in check by predators. If the predators are removed by hunting or poisoning, the prey population may increase to a point at which it reaches the carrying capacity, or may even increase beyond the carrying capacity to a point at which crucial resources are overtaxed and the population crashes.

Community composition is also affected by **mutualistic relationships** (a type of symbiosis), in which two species benefit each other.

Mutualistic species reach higher densities when they occur together than when only one of the species is present. Common examples of such mutualisms are plants with fleshy fruits and fruit-eating birds that disperse their seeds; flower-pollinating insects and flowering plants; the fungi and algae that together form lichens; plants that provide homes for ants, which supply them with nutrients (Figure 1.6); and corals and the algae that live inside them (Bawa 1990; Buchmann and Nabhan 1996). At its extreme, mutualism involves two species that are always found together and apparently cannot survive without each other. For example, the deaths of certain types of coral-inhabiting algae may be followed by the weakening and subsequent deaths of their associated coral species.

Trophic levels. Species in a biological community can be classified according to how they obtain energy from the environment (Figure 1.7). These classes are called **trophic levels**, and the first of them comprises the **photosynthetic species** (also known as **primary producers**), which obtain energy directly from the sun to build the organic molecules they need to live and grow. In terrestrial environments, higher plants such as flowering plants, gymnosperms, and ferns are responsible for most photosynthesis, while in aquatic environments, seaweeds, single-celled algae, and cyanobacteria (also called blue-green algae) are the most important primary producers.

(A)

(B)

1.6 A mutualistic relationship. (A) This *Myrmecodia* in Borneo is an epiphyte—a plant that grows on the surface of another plant. *Myrmecodia* produces a tuber at its base that is filled with hollow chambers, as seen in (B). The chambers are occupied by ant colonies, which use some chambers as nesting sites and some as "dumps" for wastes and dead ants. The epiphyte absorbs the mineral nutrients it needs for growth from these "dumps," while the ants obtain a safe nest. In the epiphyte–tree relationship shown in (A), the epiphyte benefits while the tree it grows on is neither benefited nor harmed. (Photographs by R. Primack.)

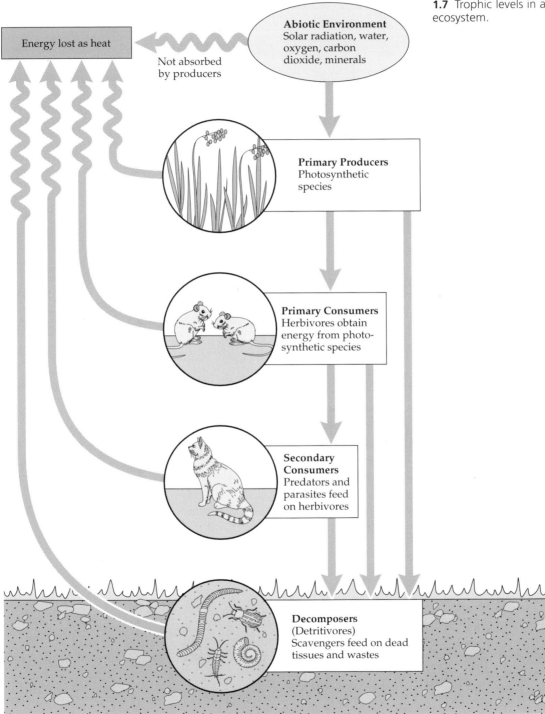

1.7 Trophic levels in a field ecosystem.

Energy lost as heat

Not absorbed by producers

Abiotic Environment
Solar radiation, water, oxygen, carbon dioxide, minerals

Primary Producers
Photosynthetic species

Primary Consumers
Herbivores obtain energy from photosynthetic species

Secondary Consumers
Predators and parasites feed on herbivores

Decomposers
(Detritivores)
Scavengers feed on dead tissues and wastes

The second trophic level consists of **herbivores** (also known as **primary consumers**), which eat photosynthetic species. **Carnivores** (also known as **secondary consumers** or **predators**), which make up the third trophic level, eat other animals. **Primary carnivores** (such as foxes) eat herbivores (such as rabbits), while **secondary carnivores** (such as bass) eat other carnivores (such as frogs). Carnivores usually are predators, though some combine direct predation with scavenging behavior, and others, known as **omnivores**, include a substantial proportion of plant foods in their diets. In general, predators occur at lower densities than their prey.

    Parasites form an important subclass of predators. Parasites, such as mosquitoes, ticks, intestinal worms, and disease-causing microparasites (bacteria, viruses, and protozoans), are typically smaller than their prey, known as **hosts**, and do not kill their prey immediately. The effects of parasites range from imperceptibly weakening their hosts to totally debilitating or even killing them over time. Parasites are often important in controlling the density of their host species. When host populations are at a high density, as might occur in a confined zoo population, parasites can readily spread from one host individual to the next, causing an intense local infestation of the parasite and a subsequent decline in host density.

    Detritivores (also known as **decomposers**) are species that feed on dead plant and animal tissues and wastes, breaking down complex tissues and organic molecules. Detritivores release minerals such as nitrogen and phosphorus back into the environment, where they can be taken up again by plants and algae. The most important detritivores are fungi and bacteria, but a wide range of other species plays a role in breaking down organic materials. For example, vultures and other scavengers tear apart and feed on dead animals, dung beetles bury and feed on animal dung, and worms break down fallen leaves and other organic matter. If detritivore species were not present or were less abundant, mineral nutrients would be less available in the soil and plant growth would decline greatly.

    As a general rule, the greatest **biomass** (living weight) in an ecosystem will be that of the primary producers. In any community there are likely to be more individual herbivores than primary carnivores, and more primary carnivores than secondary carnivores. Although species can be organized into these general trophic levels, their actual requirements or feeding habitats within the trophic levels may be quite restricted. For example, a certain aphid species may feed only on one type of plant, and a certain lady beetle species may feed only on one type of aphid. These specific feeding relationships have been termed **food chains**. Such species-specific ecological requirements are an important reason for the inability of many species to increase in abundance within a community. The more common situation in many biological communities, however, is for a species to feed on several items at the trophic level below it, to compete for food with several species at its own trophic level, and in turn to be preyed upon by several species at the trophic level above it. Consequently, a more accurate description of the organization of biological communities is a **food web,** in which species are linked together through complex feeding relationships (Figure 1.8). Species at the same trophic level that use approximately the same resources in the environment are considered to be members of a **guild** of competing species.

Keystone species and resources. Within biological communities, certain species may be important in determining the ability of large numbers of other species to persist in the community. These **keystone species** affect the organization of the community to a far

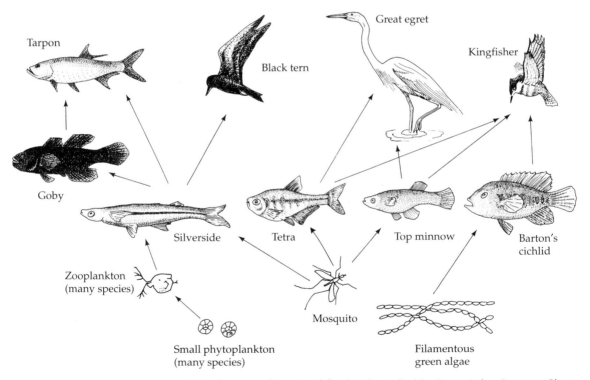

1.8 A diagram of an actual food web studied in Gatun Lake, Panama. Phytoplankton ("floating plants") such as green algae are the primary producers at the base of the web. Zooplankton are tiny, often microscopic, floating animals; they are primary consumers, not photosynthesizers, but they, along with insects and algae, are crucial food sources for fishes in aquatic ecosystems. (After G. H. Orians.)

greater degree than one would predict based only on their numbers of individuals or biomass (Terborgh 1976; Power et al. 1996). Protecting keystone species is a priority for conservation efforts because if a keystone species is lost from a conservation area, numerous other species might be lost as well (Figure 1.9). Top predators, such as wolves, are among the most obvious keystone species because they are often important in controlling herbivore populations. Without wolves, populations of deer and other herbivore species often increase, leading to overgrazing, loss of plant cover, loss of associated insect species, and soil erosion. In tropical forests, fig trees are considered to be keystone species, providing populations of many birds and mammals with a steady supply of fig fruits to eat when other preferred food is unavailable. Beavers may be considered as keystone species because they create wetland habitats used by many other species. Disease-causing organisms and parasites are other examples of keystone species that limit the densities of their host species.

Bats known as "flying foxes," which belong to the family Pteropodidae, are a classic example of a keystone species. These bats are the primary pollinators and seed dispersers of numerous economically important tree species in the Old World tropics and Pacific Islands (Cox et al. 1991). When bat colonies are overharvested by hunters, or when the trees the bats roost in are cut down, the bats' populations decline. As a result, many of the tree species in the

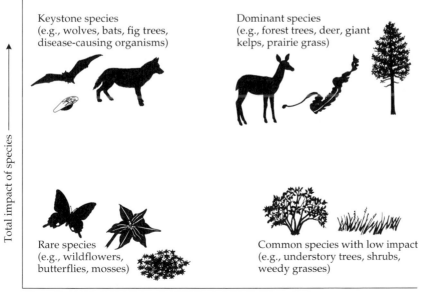

1.9 Keystone species, such as wolves, fig trees, bats, and disease-causing organisms, make up only a small proportion of the total biomass of a biological community, yet they have a huge impact on the community's organization and survival. (After Power et al. 1996.)

Keystone species
(e.g., wolves, bats, fig trees, disease-causing organisms)

Dominant species
(e.g., forest trees, deer, giant kelps, prairie grass)

Rare species
(e.g., wildflowers, butterflies, mosses)

Common species with low impact
(e.g., understory trees, shrubs, weedy grasses)

Total impact of species

Proportional biomass of species

remaining forest fail to reproduce. In short, the elimination of a sin-gle keystone species in a forest, even one that constitutes only a minute portion of the community biomass, can create a series of linked extinction events known as an **extinction cascade**, which results in a degraded ecosystem with much lower biological diver-sity at all trophic levels. Returning the keystone species to the com-munity may not necessarily restore the community to its original condition if other component species have been lost and aspects of the physical environment, such as the soil, have been damaged.

The identification of keystone species has several important implications for conservation biology. First, as we have seen, the elimination of a keystone species from a community may precipi-tate the loss of many other species (Sæther 1999). Second, in order to protect a species of particular interest, the keystone species on which it depends (either directly or indirectly) may need to be pro-tected as well. Third, if the few keystone species of a community can be identified, they can be carefully protected or even encour-aged if the area is being altered by human activity such as grazing, logging, or residential development.

In addition to the importance of keystone species to a biological community, certain resources may play key roles as well. Nature reserves are typically compared and valued in terms of size be-cause, in general, larger reserves contain more species than smaller reserves. However, area alone may not be as significant as the range of habitats and resources that the reserve contains. Particular nature reserves may contain critical **keystone resources** that occupy only a small portion of the protected area yet are crucial to many species in the community. We have already discussed the importance of clay licks to macaws. In addition, salt licks and mineral pools may pro-vide essential minerals for wildlife, particularly in inland areas with heavy rainfall. Deep pools in streams and springs may be the only refuge for fish and other aquatic species during the dry season, when water levels drop. These pools may also be the only source of drinking water for terrestrial animals for a considerable distance. Hollow tree trunks, also, are needed as breeding sites for many birds and mammal species. When old-growth forests are cleared for tree plantations, old, hollow trees that could provide breeding sites are lost, and many species may fail to breed, even though the area is still forest.

Measuring biological diversity

In addition to the definition of biological diversity generally ac-cepted by conservation biologists, there are many other specialized, quantitative definitions of biological diversity that have been devel-oped as a means of comparing the overall diversity of different communities at different geographic scales (Hellmann and Fowler

1999). These definitions have also been used to test the theory that increasing levels of diversity lead to increasing levels of community stability, productivity, and resistance to invasion by exotic species (Pimm 1991; Tilman 1999). The number of species in a single community is usually described as **species richness** or **alpha diversity**, and can be used to compare the number of species in different geographical areas or biological communities. The term **beta diversity** refers to the degree to which species composition changes along an environmental or geographical gradient. Beta diversity is high, for example, if the species composition of moss communities changes substantially on adjacent peaks of a mountain range, but beta diversity is low if most of the same species occupy the whole mountain range. **Gamma diversity** applies to larger geographical scales; it refers to the number of species in a large region or on a continent.

We can illustrate the three types of diversity with a theoretical example of three mountain ranges (Figure 1.10). Region 1 has the highest alpha diversity, with a greater mean number of species per mountain (6 species) than the other two regions. Region 2 has the highest gamma diversity, with a total of 10 species. Region 3 has a higher beta diversity (3.0) than region 2 (2.5) or region 1 (1.2) because all of its species are found on one mountain each. In practice, these three indices are often highly correlated. The plant communities of the Amazon, for instance, show high levels of diversity at the alpha, beta, and gamma scales (Gentry 1986). These quantita-

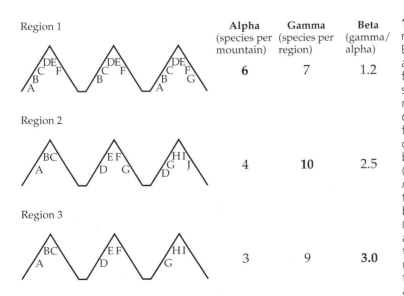

1.10 Biodiversity indices for three regions, each with three mountains. Each letter represents a population of a species. Some species are only found on one mountain, while other species are found on two or three mountains. Alpha, beta, and gamma diversity are shown for each region. If funds were available to protect only one mountain *range*, region 2 should be selected because it has the greatest (total) diversity. However, if only one *mountain* can be protected, a mountain in region 1 should be selected because these have the highest alpha (local) diversity, that is, the greatest average number of species per mountain. Each mountain in region 3 has a more distinct assemblage of species than those in the other two regions, as shown by the higher beta diversity. Overall, region 3 would be a lower conservation priority.

tive definitions of diversity are used primarily in the technical ecological literature and capture only part of the broad definition of biological diversity used by conservation biologists. However, they are useful for talking about patterns of species distribution, and for highlighting areas that have high diversity and require conservation protection.

Where Is Biological Diversity Found?

The most species-rich environments appear to be tropical rain forests, coral reefs, large tropical lakes, and the deep sea (WCMC 1992; Heywood 1995). There is also an abundance of species in tropical dry habitats, such as deciduous forests, shrublands, grasslands, and deserts (Mares 1992), and in temperate shrublands with Mediterranean climates, such as those found in South Africa, southern California, and southwestern Australia. In tropical rain forests, this diversity is primarily due to the great abundance of animal species in a single class: the insects. In coral reefs and the deep sea, the diversity is spread over a much broader range of phyla and classes (Grassle et al. 1991). Diversity in the deep sea may be due to the great age, enormous area, and stability of that environment, as well as specialization of particular sediment types (Waller 1996). The great diversity of fishes in large tropical lakes and the presence of unique species on islands is due to **evolutionary radiation** (Box 1.2) in a series of isolated, productive habitats (Kaufman and Cohen 1993).

For almost all groups of organisms, species diversity increases toward the Tropics (Huston 1994). For example, Thailand has 251 species of mammals, while France has only 93 species, despite the fact that the two countries have roughly the same land area (Table 1.2). The contrast is particularly striking in the case of trees and other flowering plants: 10 hectares of forest in Amazonian Peru or lowland Malaysia might have 300 or more species growing as trees, while an equivalent forest in temperate Europe or the United States would probably contain 30 species or fewer. Patterns of diversity in terrestrial species are paralleled by patterns in marine species, with a similar increase in species diversity toward the Tropics. For example, the Great Barrier Reef, off Australia, has 50 genera of reef-building corals at its northern end, where it approaches the Equator, but only 10 genera at its southern end.

The greatest diversity of species is found in tropical forests. Even though tropical forests occupy only 7% of the Earth's land area (see Figure 2.9), they contain over half of the world's species (Whitmore 1990). This estimate is based to some degree on only limited sampling of insects and other arthropods, groups that are thought to contain the majority of the world's species. Estimates of

Box 1.2 **The Origin of New Species**

The process whereby one original species evolves into one or more new, distinct species, known as **speciation**, was first described by Charles Darwin and Alfred Russel Wallace more than 100 years ago.

The theory of evolution of new species through **natural selection** is both simple and elegant. Individuals within a population show variations in certain characteristics, and some of these characteris-

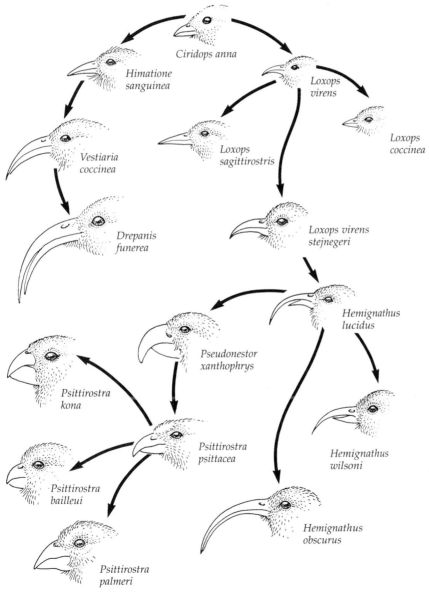

One of the most spectacular examples of adaptive radiation is displayed by the honeycreeper family. This family of birds, endemic to the Hawaiian Islands, is thought to have arisen from a single pair of birds that arrived on the islands by chance. The shapes and sizes of the bills are related to the foods eaten by each species: Long bills are for feeding on nectar; short, thick bills are for cracking seeds; and short, sharp bills are for eating insects. (After Futuyma 1986.)

Box 1.2   *(continued)*

tics are inherited—they are passed genetically from parents to offspring. These genetic variations are caused by spontaneous changes in the chromosomes and by the rearrangement of chromosomes during sexual reproduction. Differences in genetic characteristics will enable some individuals to grow, survive, and reproduce better than other individuals, an idea often characterized as "survival of the fittest." As a result of the improved survival ability provided by a certain genetic characteristic, the individuals with that characteristic will be more likely to produce offspring than individuals without it; over time, the genetic composition of the population will change.

The gene pool of a population will often change over time as the environment of the species changes. These changes may be biological (altered food availability, competitors, prey) as well as physical (changes in climate, water availability, soil characteristics). When a population has undergone so much genetic change that it is no longer able to interbreed with the original species from which it was derived, it can be considered a new species. This process, whereby one species is gradually transformed into another species, is termed **phyletic evolution**.

For two or more new species to evolve from one original ancestor, there usually must be a geographical barrier that prevents the movement of individu-

als between the various populations. For terrestrial species, these barriers may be rivers, mountain ranges, or oceans that the species cannot readily cross. Speciation is particularly rapid on islands. Island groups such as the Galápagos and Hawaiian archipelagoes contain many species-rich insect and plant genera that were originally populations of a single invading species. These populations adapted genetically to the local conditions of isolated islands, mountains, or valleys, and have diverged sufficiently from the original species to now be considered separate species. These species will remain reproductively isolated from one another even if their ranges should once again overlap. This process of local adaptation and subsequent speciation is known as **evolutionary radiation** or **adaptive radiation**. Although phyletic evolution has no net effect on biodiversity, adaptive radiation results in much greater diversity.

The origin of new species is normally a slow process, taking place over hundreds, if not thousands, of generations. The evolution of higher taxa, such as new genera and families, is an even slower process, typically lasting hundreds of thousands or even millions of years. In contrast, human activities are destroying in only a few decades the vast numbers of species built up by these slow natural processes.

the number of undescribed insect species in tropical forests range from 5 million to 30 million (May 1992); 10 million is considered a reasonable working estimate at present. If the 10 million figure is correct, it would mean that insects found in tropical forests may constitute over 90% of the world's species. About 40% of the world's flowering plant species are found in tropical forests, while 30% of the world's bird species are dependent on tropical forests (Diamond 1985).

Coral reefs constitute another concentration of species. Colonies of tiny coral animals (Figure 1.11) build the large coral reef ecosystems that are the marine equivalent of tropical rain forests in their species richness and complexity. The world's largest coral reef is the Great Barrier Reef off the east coast of Australia, which has an area of 349,000 km^2. The Great Barrier Reef contains over 300 species of corals, 1500 species of fish, 4000 species of mollusks, and 5 species

TABLE 1.2   Number of mammal species in selected tropical and temperate countries paired for comparable size

Tropical country	Area (1000 km²)	Number of mammal species	Temperate country	Area (1000 km²)	Number of mammal species
Brazil	8456	394	Canada	9220	139
DRC[a]	2268	415	Argentina	2737	258
Mexico	1909	439	Algeria	2382	92
Indonesia	1812	515	Iran	1636	140
Colombia	1039	359	South Africa	1221	247
Venezuela	882	288	Chile	748	91
Thailand	511	251	France	550	93
Philippines	298	166	United Kingdom	242	50
Rwanda	25	151	Belgium	30	58

Source: Data from WRI 1994.

[a]Democratic Republic of the Congo.

of turtles, and it provides breeding sites for some 252 species of birds (IUCN/UNEP 1988). The Great Barrier Reef contains about 8% of the world's fish species even though it occupies only 0.1% of the ocean's total surface area.

Patterns of species richness are also affected by local variation in topography, climate, environment, and geological age (Currie 1991;

1.11 Coral reefs are built up from the skeletons of billions of tiny individual animals. The intricate coral landscapes create a habitat for many other marine species, such as these French grunts shoaling near Elkhorn coral off Little Cayman, British West Indies. (Photograph © David Wrobel/Biological Photo Service.)

Huston 1994). In terrestrial communities, species richness tends to increase with lower elevation, increasing solar radiation, and increasing precipitation. Species richness can also be greater where there is a complex topography that allows genetic isolation, local adaptation, and speciation to occur over long periods of time. For example, a sedentary species occupying a series of isolated mountain peaks may eventually evolve over time into several different species, each adapted to its local mountain environment. Areas that are geologically complex produce a variety of soil conditions with very sharp boundaries between them, leading to a variety of communities and species adapted to one soil type or another. Among temperate communities, great plant species richness is found in southwestern Australia, South Africa, and other areas with a Mediterranean climate of mild, moist winters and hot, dry summers. The shrub and herb communities in these areas are apparently rich in species due to their combination of considerable geological age and complexity of site conditions. The greatest species richness in open ocean communities exists where waters from different biological communities overlap, but the locations of these boundary areas are often unstable over time (Angel 1993).

How many species exist worldwide?

Any strategy for conserving biological diversity requires a firm grasp of how many species exist and how those species are distributed. At present, about 1.5 million species have been described. At least twice this number of species remain undescribed, primarily insects and other arthropods in the Tropics (Figure 1.12A; May 1992). Our knowledge of species numbers is imprecise because inconspicuous species have not received their proper share of taxonomic attention. For example, spiders, nematodes, and fungi living in the soil and insects living in the tropical forest canopy are small and difficult to study (Figure 1.12B). These poorly known groups could number in the hundreds of thousands, or even millions, of species. Bacteria are also very poorly known (Hammond 1992). Only about 4000 species of bacteria are recognized by microbiologists because of the difficulty of growing and identifying specimens. However, work in Norway analyzing bacterial DNA hints that there may be more than 4000 species in a single gram of soil, and an equally large number of species in marine sediments (Giovannoni et al. 1990; Ward et al. 1990). Such high diversity in small samples suggests that there could be thousands or even millions of undescribed bacteria species. Recent investigations have tried to determine if there are smaller numbers of common microbial species, or larger numbers of species with regional or local distribution (Finlay and Clarke 1999).

A lack of collecting has particularly hampered our knowledge of the number of species found in the marine environment. The marine

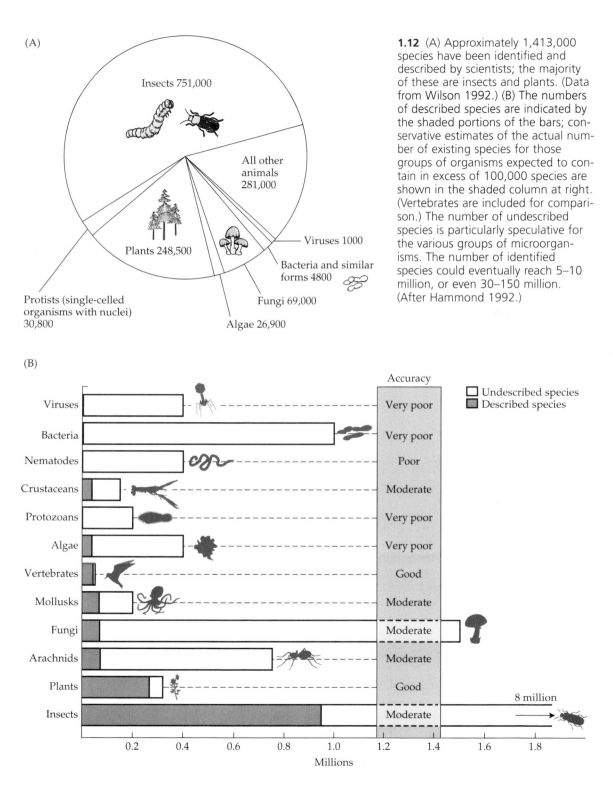

(A)

Insects 751,000

All other animals 281,000

Plants 248,500

Viruses 1000

Bacteria and similar forms 4800

Protists (single-celled organisms with nuclei) 30,800

Fungi 69,000

Algae 26,900

1.12 (A) Approximately 1,413,000 species have been identified and described by scientists; the majority of these are insects and plants. (Data from Wilson 1992.) (B) The numbers of described species are indicated by the shaded portions of the bars; conservative estimates of the actual number of existing species for those groups of organisms expected to contain in excess of 100,000 species are shown in the shaded column at right. (Vertebrates are included for comparison.) The number of undescribed species is particularly speculative for the various groups of microorganisms. The number of identified species could eventually reach 5–10 million, or even 30–150 million. (After Hammond 1992.)

(B)

Accuracy

☐ Undescribed species
■ Described species

	Accuracy
Viruses	Very poor
Bacteria	Very poor
Nematodes	Poor
Crustaceans	Moderate
Protozoans	Very poor
Algae	Very poor
Vertebrates	Good
Mollusks	Moderate
Fungi	Moderate
Arachnids	Moderate
Plants	Good
Insects	Moderate

8 million

0.2 0.4 0.6 0.8 1.0 1.2 1.4 1.6 1.8

Millions

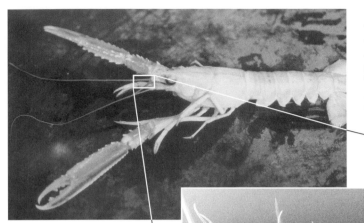

1.13 A new phylum, the Cycliophora, was first described in 1995. The phylum contains one vase-shaped species, *Symbion pandora* (around 40 of which are shown below), which attaches itself on the mouthparts of the Norway lobster, *Nephrops norvegicus* (top). (Photographs courtesy of Reinhardt Kristensen, University of Copenhagen.)

environment appears to be a great frontier of biological diversity. An entirely new animal phylum, the Loricifera, was first described in 1983 based on specimens from the deep seas (Kristensen 1983), and another new phylum, the Cycliophora, was first described in 1995 based on tiny, ciliate creatures found on the mouthparts of the Norway lobster (Figure 1.13; Funch and Kristensen 1995). In 1999, the world's largest bacteria was discovered off the Namibian coast, with individual cells as big as the eyes of fruit flies (Schulz et al. 1999). There are undoubtedly many more marine species to be discovered.

Completely new biological communities are still being discovered, often in localities that are extremely remote and inaccessible to humans. Specialized exploration techniques, particularly in the deep sea and the forest canopy, have revealed unusual community structures:

• Diverse communities of animals, particularly insects, are adapted to living in the canopies of tropical trees, rarely if ever descending to the ground (Wilson 1991; Moffat 1994). Canopy towers and walkways are now allowing researchers access to this environment.

- In a remote, mountainous rain forest reserve on the border between Vietnam and Laos, biologists were amazed to discover three species of large mammals new to science, now known as the giant muntjac, the Vu Quang ox, and the slow-running deer (Linden 1994).

- The floor of the deep sea, which remains largely unexplored due to the technical difficulties of transporting equipment and people under high water pressure, has unique communities of bacteria and animals that grow around deep-sea geothermal vents (Tunnicliffe 1992). Undescribed, active bacteria have even been found in marine sediments up to 500 meters below the sea floor, where they undoubtedly play a major chemical and energetic role in this vast ecosystem (Parkes et al. 1994).

- Recent drilling projects have discovered diverse bacterial communities living up to 2.8 kilometers beneath the Earth's surface, at densities of up to 100 million bacteria per gram of rock. These communities are being actively investigated as a source of novel chemicals, for their potential usefulness in degrading toxic chemicals, and for insights into whether life could exist on other planets (Fredrickson and Onstatt 1996; Fisk et al. 1998).

Extinction and Economics: Losing Something of Value

To discover, catalogue, and preserve the great diversity of species, a new generation of conservation biologists must be trained, and an increased priority must be given to the museums, universities, conservation organizations, and other institutions that support this work. Such a change will require a significant shift in current political and social thinking; governments and communities throughout the world must realize that biological diversity is extremely valuable—indeed, essential—to human existence. Ultimately, change will occur only if people feel that by continuing to damage biological communities they are truly losing something of value. But what is it we are losing? Why should anyone care if a species becomes extinct? What, precisely, is so terrible about extinction?

Patterns of extinction

The diversity of species found on the Earth has been increasing since life first originated. This increase has not been steady, but rather has been characterized by periods of high rates of speciation, followed by periods of minimal change and five past episodes of **mass extinction** (Wilson 1989; Raup 1992). The most massive

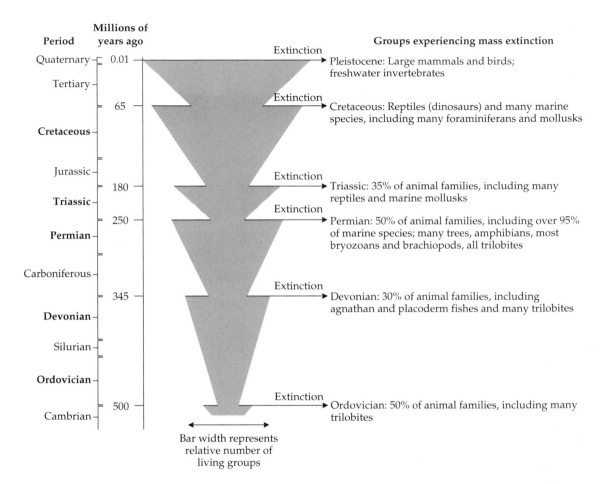

1.14 Although the total number of families and species on Earth has increased over the eons, during each of five episodes of natural mass extinction (names in bold at left), a large percentage of these groups disappeared. The most dramatic period of loss occurred about 250 million years ago, at the end of the Permian period. We are now at the start of a sixth episode, the Pleistocene extinction, characterized by the elimination of species by human populations through habitat loss, overharvesting, and the impact of invasive species.

extinction took place at the end of the Permian period, 250 million years ago, when 77%–96% of all marine animal species are estimated to have become extinct (Figure 1.14). It is quite likely that some massive perturbation, such as widespread volcanic eruptions or a collision with an asteroid, caused such a dramatic change in the Earth's climate that many species were no longer able to exist. It took the process of evolution about 50 million years to regain the number of families lost during the Permian mass extinction. Species extinctions occur even in the absence of violent disturbance, however. One species may outcompete another or drive it to extinction through predation. A successful species may evolve into another in response to environmental changes or due to random changes in its gene pool. The factors determining the persistence or failure of a particular species are not entirely clear, but extinction is as much a part of the natural cycle as speciation is.

If extinction is a part of natural processes, why is the loss of species of concern? The answer lies in the relative rates of extinction and speciation. Speciation is typically a slow process, occurring through the gradual accumulation of mutations and shifts in allele frequencies over thousands, if not millions, of years. As long as the rate of speciation equals or exceeds the rate of extinction, biodiversity will remain constant or increase. In past geological periods, the loss of existing species was eventually balanced or exceeded by the evolution of new species. However, current rates of extinction are 100 to 1000 times those of past rates. This current episode of extinctions, sometimes called the **sixth extinction**, is due almost exclusively to human activities. This loss of species is unprecedented, unique, and irreversible.

Ecological economics

Before the trend of species extinctions can be reversed, its fundamental causes must be understood. What factors cause humans to act in a destructive manner? Ultimately, environmental degradation occurs for economic reasons. Forests are logged for revenue from timber sales. Animals are hunted for their meat, fur, and other products. Natural habitats are converted into cropland because people have nowhere else to farm. Species are introduced onto new continents and islands without any consideration for the resulting environmental devastation. Because the underlying cause of environmental damage is so often economic in nature, the solution must incorporate economic principles as well. Conservation biologists are increasingly incorporating economic elements into their research programs, and economic arguments into their recommendations.

An understanding of a few fundamental economic principles will serve to clarify why people treat the environment in a shortsighted, wasteful manner. One of the most universally accepted tenets of modern economic thought is that a voluntary transaction takes place only when it is beneficial to both parties involved. A baker who sells his loaves for fifty dollars apiece will find few customers. Likewise, a customer who is willing to pay only five cents for a loaf will soon go hungry. A transaction will occur between the baker and the bread buyer only when a mutually agreeable price is set that benefits both parties. Adam Smith, an eighteenth-century philosopher whose writings are the foundation of much of modern economic thought, stated "It is not upon the benevolence of the butcher, the baker, or the brewer that we eat our daily bread, but upon his own self-interest" (*Wealth of Nations*, 1776). All parties involved in an exchange have the expectation of improving their own situation. The sum of each individual acting in his or her own self-interest results in society as a whole becoming wealthier and more prosperous: It is as if an "invisible hand" is guiding the market.

It is beyond the scope of this text to address all the assumptions in and exceptions to this principle of free exchange benefiting society. However, there is one notable exception that directly affects environmental issues. It is generally assumed that the costs and benefits of free exchange are accepted and borne by the participants in the transaction. In some cases, however, certain costs are incurred, or benefits enjoyed, by individuals not directly involved in the exchange. These outside costs or benefits are known as **externalities**. Perhaps the most notable and frequently overlooked externality is the environmental damage that occurs as an indirect consequence of human economic activity. Where externalities exist, the market fails to provide solutions that result in a more prosperous society. This **market failure** results in a misallocation of resources that may allow certain individuals to benefit at the expense of the environment and the entire society.

The fundamental challenge facing the conservation biologist is to ensure that all the costs of a transaction are understood as well as the benefits. Companies or individuals involved in activities that result in ecological damage generally do not pay the full cost of such damage, and often pay nothing at all. Instead, the environmental costs are often borne by people living nearby who obtain little or no benefit from these activities.

For example, an oil refinery that emits toxic fumes and liquid wastes benefits from the sale of fuel, as does the consumer of the product. Yet the costs of this transaction—decreased air quality, a polluted water supply, and increased respiratory disease—are distributed widely throughout society, often strongly affecting people living in the local area. Such examples of market failure can also operate in international trade: The price of fancy, farm-raised shrimp grown in countries of the developing world and sold in wealthy industrial countries does not include the cost of environmental damage to coastal wetlands and its effects on nearby villages that were caused when the land was cleared to build the shrimp ponds.

In response to this challenge to account for all costs in economic transactions, including environmental costs, a discipline is being developed that integrates economics, environmental science, ecology, and public policy, and includes valuations of biological diversity in economic analyses (Barbier et al. 1994; Masood and Garwin 1998). This discipline is known as **ecological economics**. Conservation biologists are increasingly using the concepts and vocabulary of ecological economics because government officials, bankers, and corporate leaders can be more readily convinced of the need to protect biological diversity if provided with an economic justification for doing so.

The environmental costs of large projects, such as dams, roads, irrigation systems, and plantation agriculture, are increasingly being calculated in the form of **environmental impact assessments** that consider the present and future effects of the projects on the environment. The environment is often broadly defined to include not just harvestable natural resources, but also air and water quality, the lives of local people, and endangered species. In its most comprehensive form, a **cost–benefit analysis** compares the values gained by a project against the costs of the project and the values lost (Perrings 1995). In theory, if the analysis shows that a project is profitable, it should go forward, while if the project is unprofitable, it should be stopped. In practice, cost–benefit analyses are often only rough approximations, because the valuations of the benefits and costs may be difficult to assign and may change over time. Ecological economists are contributing to these analyses by including the full range of costs and benefits.

Attempts have been made to include the loss of natural resources in calculations of the widely used Gross Domestic Product (GDP) and other indices of national well-being (Daly and Cobb 1989; Repetto 1992). The problem with the GDP as it is usually calculated is that it measures all economic activity in a country, not just environmentally beneficial activity. Nonsustainable and unproductive economic activities (including overfishing of coastal waters and poorly managed strip mining) cause the GDP to increase, even though these activities are actually unrelated or destructive to the long-term well-being of the country. Even environmental disasters such as the Exxon Valdez oil spill and the Gulf War appear as increases in the GDP because of the temporary increases in jobs and purchases needed for the cleanups. In actuality, the economic costs to a country associated with environmental damage can be considerable and often offset the gains the country apparently attains through agricultural and industrial development. In Costa Rica, for example, the value of the forests destroyed during the 1980s greatly exceeds the income produced from forest products, so that the forestry sector actually represented a drain on the wealth of the country (Repetto 1992).

Common property resources

Many natural resources, such as clean air, clean water, soil quality, rare species, and even scenic beauty are considered to be **common property resources** that are owned by society at large. These resources are often not assigned a monetary value. People, industries, and governments use—and damage—these resources without paying more than a minimal cost, and sometimes paying nothing at all, a situation described as **the tragedy of the commons** (Hardin 1985). In the more complete systems of "green" accounting being developed, such as National Resource Accounting, the use of such common property resources is included as part of the internal cost of doing business, instead of being regarded as an externality. When people and organizations have to pay for their actions, they will do less damage to the environment. Some suggestions for bringing this about include higher taxes on fossil fuels, penalties for inefficient energy use and pollution, and mandatory recycling programs. Also, subsidies to industries and activities that pollute the environment could be eliminated. As a result, new investment could be redirected to activities that improve the environment or are at least benign, and that provide the most benefits to the greatest number of people, particularly those living in poverty. Finally, financial penalties for damaging biological diversity could be developed and made so severe that industries would be more careful of the natural world.

Demonstrating the value of biodiversity and natural resources is a complex matter because value is determined by a variety of economic and ethical factors. A major goal of ecological economics is to develop methods for valuing the components of biological diversity. A number of approaches have been developed to assign economic values to genetic variability, species, communities, and ecosystems. One of the most useful is that used by McNeely et al. (1990) and Barbier et al. (1994). In this framework, values are divided between **direct values** (known in economics as *private goods*), which are assigned to those products harvested by people, such as fish, timber, and medicinal plants, and **indirect values** (in economics, *public goods*), which are assigned to benefits provided by biological diversity that do not involve harvesting or destroying the resource. Benefits that can be assigned indirect value include water quality, soil protection, recreation, education, scientific research, and regulation of climate. Biological diversity also has **option value** for its ability to provide new goods and services in the future, and **existence value** based on how much people are willing to pay to protect a species from going extinct or a particular biological community from being destroyed.

Direct Economic Values

Direct values are assigned to those products that are directly harvested and used by people. These values can often be readily calculated by observing the activities of representative groups of people, by monitoring collection points for natural products, and by examining import and export statistics. Direct values can be further divided into **consumptive use value**, for goods that are consumed locally, and **productive use value**, for products that are sold in markets.

Consumptive use value

Consumptive use value can be assigned to goods such as fuelwood and wild animals that are consumed locally and do not appear in the national and international marketplaces. People living close to the land often derive a considerable proportion of the goods they require for their livelihood from the environment around them. These goods do not typically appear in the GDP of countries because they are typically neither bought nor sold, or are sold in only local markets. However, if rural people are unable to obtain these products, as might occur following environmental degradation, overexploitation of natural resources, or even the creation of a protected reserve, then their standard of living will decline, possibly to the point where they are unable to survive and must move to another location.

Studies of traditional societies in the developing world show how extensively these people use their natural environment to supply their needs for fuelwood, vegetables, fruit, meat, medicinal plants, cordage, and building materials (Figure 1.15; Myers 1994; Balick and Cox 1996). For example, about 80% of the world's population still relies on traditional medicines derived from plants and animals as their primary source of medical treatment (Farnsworth 1988; Tuxill 1999). Over 5000 species are used for medicinal purposes in China, while 2000 species are used in the Amazon basin (Schultes and Raffauf 1990).

1.15 Natural products are critical to the lives of people throughout the world. (A) One of the most important natural products required by local people is fuelwood, particularly in Africa and southern Asia. Here a woman in Burkina Faso gathers kindling. (World Bank Photo by Yosef Hadar, © IBRD.) (B) A wide variety of plants and other natural products are used in Chinese medicine. In this shop, a pharmacist is preparing and packaging traditional medicines. Note the mortar and pestle used to grind the ingredients into powder, and the abacus used to calculate prices. (Photograph © Catherine Pringle/ Biological Photo Service.)

(A)

(B)

One of the crucial requirements of rural people is protein, which they often obtain by hunting wild animals for meat. In many areas of Africa, wild game constitutes a significant portion of the protein in the average person's diet: in Botswana, about 40%, and in the Democratic Republic of the Congo, 75% (Myers 1988b). Throughout the world, 108 million tons of fish, crustaceans, and mollusks, mainly wild species, are harvested each year, with 91 million tons constituting marine catch and 17 million tons constituting freshwater catch (WRI 1998). Much of this catch is consumed locally.

Consumptive use value can be assigned to a product by considering how much people would have to pay to buy equivalent products in the market if the local sources were no longer available. One example of this approach was an attempt to estimate the number of wild pigs harvested by native hunters in Sarawak, East Malaysia, in part by counting the number of shotgun shells used in rural areas and interviewing hunters. This pioneering (and somewhat controversial) study estimated that the consumptive use value of the wild pig meat was around $40 million per year (Caldecott 1988). But in many cases, local people do not have the money to buy products in the market. When the local resource is depleted, they are forced into rural poverty, or they migrate to urban centers.

While dependence on local natural products is primarily associated with the developing world, there are rural areas of the developed countries, such as United States and Canada, where hundreds of thousands of people are dependent on fuelwood for heating and on wild game for meat. Many of these people would be unable to survive in such remote areas if they had to buy fuel and meat.

Productive use value

Productive use value is a direct value assigned to products that are harvested from the wild and sold in commercial markets, at both the national and international levels. These products are typically valued by standard economic methods at the price that is paid at the first point of sale minus the costs incurred up to that point, rather than at the final retail cost of the products; as a result, what may appear to be minor natural products may actually be the starting points of major manufactured products (Godoy et al. 1993). For example, wild cascara (*Rhamnus purshiana*) bark gathered in the western United States is the major ingredient in certain brands of laxatives; the purchase price of the bark is about $1 million per year, but the final retail price of the medicine is $75 million (Prescott-Allen and Prescott-Allen 1986). The range of products obtained from the natural environment and then sold in the marketplace is enormous, but the major ones are fuelwood, construction timber, fish and shellfish, medicinal plants, wild fruits and vegetables, wild meat and skins, fibers, rattan, honey, beeswax, natural dyes, seaweed, animal fodder, natural perfumes, and plant gums and resins (Radmer 1996; Baskin 1997).

The productive use value of natural resources is significant, even in industrial nations. Prescott-Allen and Prescott-Allen (1986) calculated that 4.5% of the U.S. GDP depends in some way on wild species, an amount averaging about $87 billion per year. The percentage would be far higher for developing countries that have less industry and a higher percentage of the population living in rural areas.

At present, timber is among the most significant products obtained from natural environments, with a value of over $120 billion per year in international trade (WRI 1998). Timber products are being exported at a rapid level from many tropical countries to earn foreign currency, to provide capital for industrialization, and to pay foreign debt. In tropical countries such as Indonesia and Malaysia, timber products are among the top export earners, accounting for billions of dollars per year (Primack and Lovejoy 1995) (Figure 1.16A). Nonwood products from forests, including game, fruits, gums and resins, rattan, and medicinal plants, also have a great productive use value (Figure 1.16B). For example, nontimber products account for 63% of the total foreign exchange earned by India from the export of forest products (Gupta and Guleria 1982). These nontimber products, which are sometimes erroneously called "minor forest products," are in reality very important economically and may be greater in value over time than the one-time timber

1.16 (A) The timber industry is a major source of revenue in many countries. Here trees are harvested from rain forest in Indonesian Borneo. (Photograph © Wayne Lawler, Photo Researchers, Inc.) (B) Nontimber forest products are often important in local and national economies. Many rural people supplement their incomes by gathering natural forest products to sell in local markets. Here a Land Dayak family in Sarawak (Malaysia) sells wild honey and edible wild fruits. (Photograph by R. Primack.)

(A)

(B)

harvests (Panayotou and Ashton 1992; Daily 1997). The value of nontimber products, along with the value of forests in other ecosystem functions, provides a strong economic justification for maintaining forests in many areas of the world.

The greatest productive use value for many species lies in their ability to provide new founder stock for industry and agriculture and for the genetic improvement of agricultural crops (Baskin 1997). Some wild species of plants and animals that are currently harvested on a local scale can be grown on plantations and ranches, and some can be cultured in laboratories. The wild populations provide the initial breeding stock for these colonies and are a source of material for genetic improvement of the domesticated populations. In the case of crop plants, a wild species or variety might provide a particular gene that confers pest resistance or increased yield. This gene needs to be obtained from the wild only once; it then can be incorporated into the breeding stock of the crop species and stored in gene banks. The continued genetic improvement of cultivated plants is necessary not only for increased yield, but also to guard against pesticide-resistant insects and increasingly virulent strains of fungi, viruses, and bacteria. Catastrophic failures of crop plants can often be directly linked to low genetic variability: The 1846 potato blight in Ireland, the 1922 wheat failure in the Soviet Union, and the 1984 outbreak of citrus canker in Florida were all related to low genetic variability among crop plants (Plucknett et al. 1987).

Development of new varieties can also have a noticeable economic effect. Genetic improvements in crops in the United States were responsible for increasing the value of the harvest by an average of $1 billion per year from 1930 to 1980 (OTA 1987). Genetic improvements of rice and wheat varieties during the "Green Revolution" increased harvests in Asia by an estimated $3.5 billion per year (WCMC 1992). Genes for high sugar content and large fruit size from wild tomatoes in Peru have been transferred into domestic varieties of tomatoes, resulting in an enhanced value of $80 million to the industry (Iltis 1988). The discovery of a wild perennial relative of corn in the Mexican state of Jalisco (see Chapter 5) is potentially worth billions of dollars to modern agriculture because it could allow the development of a high-yielding perennial corn crop and eliminate the need for annual plowing.

Wild species can also be used as **biological control agents** (Van Driesche and Bellows 1996). Biologists have sometimes controlled exotic, noxious species by searching the pest species' original habitat for a species that limits its population. This control species can then be brought to the new locality, where it can be released to control the pest. One such example is the control of the South American cassava mealybug (*Phenacoccus manihoti*), a devastating pest of cassava (*Manihot esculenta*), a major food plant in Africa (Herren and Neuenschwander 1991). After it was accidentally introduced in Africa, the mealybug was causing an estimated damage of $2 billion per year to the cassava crop, and substantially reducing the food available to 200 million Africans. After an intensive world-wide search, entomologists discovered a tiny wasp (*Apoanagyrus lopezi*) in Paraguay that was previously unknown to science and that specifically parasitizes the cassava mealybug (Figure 1.17). A program to breed and airdrop 250,000 wasps per week over a wide area led to the effective control of the mealybugs and a 95% reduction in crop losses.

The natural world is also an important source of medicines. More than 75% of the leading 150 prescription drugs used currently in the United States were originally derived from plants, animals, fungi, and bacteria (Dobson 1998). Twenty-five percent of prescrip-

(A)

(B)

(C)

1.17 The starchy roots of cassava plants represent one of the most important food sources in tropical Africa. (A) A field of cassava plants is shown in the foreground of this photograph. (Photograph courtesy of P. Philpot, IITA.) When cassava mealybugs were introduced in Africa, they caused extensive damage to cassava plants (B). Scientists found a tiny wasp (C) that lays its eggs in the mealybug's larvae, providing an effective means of biological control. (Photographs in B and C courtesy of P. Neuenschwander, IITA.)

tion drugs used in the United States contain active ingredients derived from plants. Two potent drugs derived from the rose periwinkle (*Catharanthus roseus*) of Madagascar, for example, have proved to be effective at treating Hodgkin's disease, leukemia, and other blood cancers. Treatment using these drugs has increased the survival rate of childhood leukemia from 10% to 90%. Many of the most important antibiotics, such as penicillin and tetracycline, are derived from fungi and other microorganisms (Eisner 1991; Dobson 1995). Most recently, the fungus-derived drug cyclosporine has proved to be a crucial element in the success of heart and kidney transplants. Many other important new medicines were first identified in venomous animals such as snakes and arthropods; and marine species have also been a rich source of chemicals with valuable medical applications.

The biological communities of the world are being continually searched for new plants, animals, fungi, and microorganisms that can be used to fight human diseases such as cancer and AIDS (Figure 1.18A; Plotkin 1993; Balick and Cox 1996; Grifo and Rosenthal 1997). These searches are generally carried out by government research institutes and pharmaceutical companies. To facilitate the search for new medicines and to profit financially from new products, the Costa Rican government established the National Biodiversity Institute (INBio), which collects biological products and supplies samples to drug companies (Figure 1.18B). The Glaxo Wellcome corporation, a Brazilian biotechnology company, and the Bra-

1.18 (A) Antonio Cue of Belize carries on the traditions of his Mayan ancestors as he prepares useful medicines from plants that are growing in the local area. He is now working with scientists to determine if chemicals in these plants can be developed for use in modern medicine. (Photograph courtesy of M. J. Balick.) (B) Taxonomists at INBio sort and classify Costa Rica's rich array of plant and animal species, samples of which are sent to drug companies. (Photograph by Steve Winter.)

(A)

(B)

zilian government recently signed a $3 million contract to sample, screen, and investigate approximately 30,000 plants, fungi, and bacteria from Brazil, with part of the royalties going to support scientific research and local community-based conservation and development projects (Neto and Dickson 1999). Programs such as these provide financial incentives for countries to protect their natural resources, and to preserve the knowledge of biodiversity possessed by indigenous inhabitants.

Indirect Economic Values

Indirect values can be assigned to aspects of biological diversity, such as environmental processes and ecosystem services, that provide economic benefits without being harvested and destroyed during use. Because these benefits are not goods or services in the usual economic sense, they do not typically appear in the statistics of national economies, such as the GDP. However, they may be crucial to the continued availability of the natural products on which the economies depend. If natural ecosystems are not available to provide these benefits, substitute sources must be found, often at great expense.

Nonconsumptive use value

Biological communities provide a great variety of environmental services that are not consumed through use: They prevent flooding and soil erosion, purify water, and provide places to enjoy and study nature. This **nonconsumptive use value** can sometimes be calculated for specific services. Economists are just beginning to calculate the value of ecosystem services at regional and global levels. The calculations are still at a preliminary level, but they suggest that the value of ecosystem services is enormous, around $32 trillion per year, greatly exceeding the direct use value of biodiversity (Costanza et al. 1997). Because this amount is greater than the global gross national product of $18 trillion per year, the point could be made that human societies are totally dependent on natural systems, and could not pay to replace these ecosystem services currently provided for free if they were permanently degraded or destroyed. Other estimates of the value of biological diversity are lower (Pimental et al. 1997), and many economists are in sharp disagreement about how the calculations should be done (Masood and Garwin 1998), indicating that more work is needed on this important topic. The following is a discussion of some of the general benefits of conserving biological diversity that typically do not appear on the balance sheets of environmental impact assessments or in GDPs.

Ecosystem productivity. The photosynthetic capacity of plants and algae allows the energy of the sun to be captured in living tissue. This plant material is the starting point for innumerable food chains, leading to all of the animal products that are harvested by people. Approximately 40% of the productivity of the terrestrial environment is now used directly or indirectly by people (Vitousek 1994). Destruction of the vegetation in an area through overgrazing by domestic animals, overharvesting of timber, or frequent fires destroys the system's ability to make use of solar energy, eventually leading to a loss of production of plant biomass and the deterioration of the animal community (including humans) living at that site.

Likewise, coastal estuaries are areas of rapid plant and algal growth that provide the starting point for food chains leading to commercial stocks of fish and shellfish. The U.S. National Marine Fisheries Service has estimated that damage to coastal estuaries has cost the United States over $200 million per year in lost productive use value of commercial fish and shellfish and in lost recreational use value of fish caught for sport (McNeely et al. 1990). Even when degraded or damaged ecosystems are rebuilt or restored—often at great expense—they usually do not perform their ecosystem functions as well as before, and they almost certainly do not contain their original species composition or species richness.

Scientists are actively investigating how the loss of one or more individual species from biological communities affects ecosystem productivity (Chapin et al. 1998). Many studies of natural and experimental grassland communities are now confirming that as species are lost, overall productivity declines and the community is less flexible in responding to environmental disturbances such as drought (Tilman et al. 1996). It is a safe bet that when species are lost, biological communities will be less able to adapt to the changing conditions created by human activity, including the altered weather conditions associated with rising CO_2 levels and global climate change.

Protection of water and soil resources. Biological communities are of vital importance in protecting watersheds, buffering ecosystems against extremes of flood and drought, and maintaining water quality (Wilson and Carpenter 1999). Plant foliage and dead leaves intercept the rain and reduce its impact on the soil, and plant roots and soil organisms aerate the soil, increasing its capacity to absorb water. This increased water-holding capacity reduces the flooding that occurs after heavy rains and allows a slow release of water for days and weeks after the rains have ceased.

When vegetation is disturbed by logging, farming, and other human activities, rates of soil erosion and even landslides increase rapidly, decreasing the value of the land for human activities. Damage to the soil limits the ability of plant life to recover following disturbance and can render the land useless for agriculture. In addition, soil particles suspended in water from runoff can kill freshwater animals, coral reef organisms, and the marine life in coastal estuaries. This silt also makes river water undrinkable for human communities along the rivers, leading to a decline in human health. Increased soil erosion can lead to premature filling of the reservoirs behind dams, leading to a loss of electrical output, and may create sandbars and islands, reducing the navigability of rivers and ports.

Unprecedented catastrophic floods in Bangladesh, India, the Philippines, and Thailand have been associated with recent extensive logging in watershed areas; such incidents have led to calls by local people for bans on logging. Flood damage to India's agricultural areas has led to massive government and private tree-planting programs in the Himalayas. In the industrial nations of the world, wetlands protection is being made a priority to prevent flooding of developed areas. The conversion of floodplain habitat to farmland along the Mississippi River in the midwestern United States and along the Rhine River in Europe is considered a major factor in the massive damaging floods of recent years.

In many areas of the world, growing cities increasingly face shortages of critical water supplies needed for drinking, washing, irrigation, and industrial use. The costs of water treatment plants are so enormous that protection of existing watersheds is a high priority and provides a means of valuing ecosystem services. The need to protect water supplies led New York City to pay $1 billion to rural counties in New York state to maintain forests on the watershed surrounding its reservoirs. This was a good investment because water treatment plants doing the same job would have cost $8 to $9 billion (McKibben 1996).

Regulation of climate. Plant communities are important in moderating local, regional, and probably even global climate conditions (Clark 1992; Couzin 1999). At the local level, trees provide shade and transpire water, which reduces the local temperature in hot weather. This cooling effect reduces the need for fans and air conditioners and increases the comfort and work efficiency of people. Trees are also locally important as windbreaks and in reducing heat loss from buildings in cold weather.

At the regional level, transpiration from plants recycles water back into the atmosphere so that it can return as rain. Loss of vegetation from regions of the world such as the Amazon Basin and West Africa could result in a regional reduction of average annual rainfall (Fearnside 1990). At the global level, plant growth is tied to carbon cycles. A loss of vegetation cover results in reduced uptake of carbon dioxide by plants, contributing to the rising carbon dioxide levels that lead to global warming (Kremen et al. 1999). Plants are also the "green lungs" of the planet, producing the oxygen on which all animals depend for respiration.

Waste disposal and nutrient retention. Biological communities are capable of breaking down and immobilizing pollutants such as heavy metals, pesticides, and sewage that have been released into the aquatic environment by human activities (Odum 1997). Fungi and bacteria are particularly important in this role. The excess nutrients released in the breakdown process can then be taken up by algae and plants, beginning new food chains. The value of aquatic biological communities in waste treatment and nutrient processing and retention has been estimated at around $18 trillion per year (Costanza et al. 1997).

An excellent example of this ecosystem function is provided by the New York Bight, a 2000 square-mile (5200 km^2) bay at the mouth of the Hudson River. The New York Bight acts as a free sewage disposal system into which is dumped the waste produced by the 20 million people in the New York metropolitan area (Young et al. 1985). If the New York Bight becomes overwhelmed and damaged by a combination of sewage overload and coastal development, an alternative waste disposal system, including massive waste treatment facilities and giant landfills, will have to be developed at a cost of tens of billions of dollars.

Species relationships. Many of the species harvested by people for their productive use value depend on other wild species for their continued existence. Thus, a decline in a wild species of little immediate value to humans may result in a corresponding decline in a harvested species that is economically important. For example, the wild game and fish harvested by people are dependent on wild insects and plants for their food. A decline in insect and plant populations will result in a decline in animal harvests. Crop plants benefit from birds and predatory insects such as praying mantises (family Mantidae), which feed on pest insect species that attack the crops. Also, numerous crop species require insects for pollination and the resulting production of seeds and fruits (Buchmann and Nabhan 1996). Many useful wild plant species depend on fruit-eating animals, such as bats and birds, to disperse their seeds (Fujita and Tuttle 1991).

One of the most economically significant relationships in biological communities is that between many forest trees and crop plants and the soil organisms, such as fungi and bacteria, that provide them with essential nutrients through the breakdown of dead plant and animal matter. The poor growth and massive diebacks of trees that are occurring throughout Europe may be attributable in part to the deleterious effects of acid rain and air pollution on soil fungi (Cherfas 1991).

Recreation and ecotourism. A major focus of recreational activity is the nonconsumptive enjoyment of nature through activities such as hiking, photography, whale-watching, and bird-watching (Duffus and Dearden 1990). The monetary value of these activities, sometimes called amenity value, can be considerable (Figure 1.19). For example, 84% of Canadians participate in nature-related recreational activities that have an estimated value of US$800 million per year (Fillon et al. 1985). In the United States, almost 100 million adults and comparable numbers of children are involved each year in some form of nondestructive nature recreation, spending at least $54 billion on wildlife viewing, fishing, and hunting. More complete systems of accounting that include fees, travel, lodging, food, and equipment suggest that the recreational value of the world's ecosystems could be as high as $800 billion per year (Costanza et al. 1997).

In places of national and international significance for conservation or exceptional scenic beauty, such as Yellowstone National Park, the nonconsumptive recreational use value often dwarfs that of other local industries, including farming, mining, and logging (Power 1991). Even recreational activities such as sport hunting and fishing, which in theory are consumptive uses, could be considered nonconsumptive because the food value of the animals taken by fishermen and hunters is typically insignificant in comparison with

1.19 Most people find interacting with other species to be an educational and uplifting experience. Here a group of ecotourists greet a minke whale (*Balaenoptera acutorostrata*) that is being rescued after becoming entangled in a trawler's gill net; the float behind the whale was attached to the net to keep the whale at the surface so it could breathe. Later, rescuers were able to release the whale from the netting. Such meetings—which usually take place at greater distances, as in a more traditional "whale watch" setting or on "photo safaris" in Africa—can enrich human lives. (Photograph by Scott Kraus, New England Aquarium.)

the time and money spent on these activities. Particularly in rural economies, fishing and hunting generate hundreds of millions of dollars. The value of these recreational activities may be even greater than these numbers suggest because many park visitors, sport fishermen, and hunters indicate that they would be willing to pay even higher admission and licensing fees, if necessary.

Ecotourism is a rapidly growing industry in many countries, involving 200 million people per year and earning billions of dollars per year worldwide. Ecotourists visit a country and spend money wholly or in part to experience its biological diversity and to see particular species (Lindberg 1991; Ceballos-Lascuráin 1993). The role of ecotourism is particularly important in many tropical countries. By charging high visitor fees, Rwanda developed a gorilla tourism industry that was the country's third-largest foreign currency earner until the recent civil disturbances. Ecotourism has traditionally been a key industry in East African countries such as Kenya and Tanzania, and is increasingly important in many American and Asian countries. The value of viewing elephants alone in Kenya has been estimated to be $25 million per year (Brown 1993). Ecotourism can provide one of the most immediate justifications for protecting biological diversity, particularly when these activities are integrated into overall management plans (Wells and Brandon 1992). However, there is a danger that tourist facilities will provide a sanitized fantasy experience, rather than allowing visitors to be aware of or even see the serious social and environmental problems that endanger biological diversity (Figure 1.20). Ecotourist activities themselves can also contribute to the degradation of sensitive areas, as when tourists unwittingly trample wildflowers, break coral, dis-

1.20 Facilities for ecotourists sometimes create a tropical fantasy that disguises and ignores the realities of the problems that plague developing countries. (Cartoon from *E. G. Magazin*, Germany.)

rupt nesting bird colonies, and frighten animals away from water sources and feeding sites (Giese 1996).

Educational and scientific value. Many books, magazines, television programs, computer materials, and movies produced for educational and entertainment purposes are based on nature themes (for example, Wilson and Perlmann 1999; Morell 1999). Increasingly, natural history materials are being incorporated into school curricula. These educational programs are probably worth billions of dollars per year. A considerable number of professional scientists and educators, as well as highly motivated amateurs, are engaged in making ecological observations and preparing education materials. In rural areas, these activities often take place in scientific field stations, which are sources of employment and training for local people. While these scientific activities provide economic benefits to the areas surrounding field stations, their real value lies in their ability to increase human knowledge, enhance education, and enrich the human experience.

Environmental monitors. Species that are particularly sensitive to chemical toxins can serve as an "early warning system" for monitoring the health of the environment. Some species can even serve as substitutes for expensive detection equipment. Among the best-known indicator species are lichens, which grow on rocks and absorb chemicals in rainwater and airborne pollution (Hawksworth 1990). High levels of toxic materials kill lichens, and each lichen species has distinct levels of tolerance for air pollution. The composition of the lichen community in an area can be used as a biological indicator of its level of air pollution, and the distribution and abundance of lichens can be used to identify areas of contamination around sources of pollution, such as smelters. Aquatic filter feeders, such as mollusks, are also useful for monitoring pollution because they process large volumes of water and concentrate toxic chemicals, such as poisonous metals, PCBs, and pesticides, in their tissues.

Option value

The **option value** of a species is its potential to provide an economic benefit to human society at some point in the future. As the needs of society change, so must the methods of satisfying those needs. Often the solution to a problem lies in previously untapped animal or plant species. The extent of the search for new natural products is wide-ranging. Entomologists search for insects that can be used as biological control agents, microbiologists search for bacteria that can assist in biochemical manufacturing processes, and zoologists are identifying species that can produce animal protein more efficiently and with less environmental damage than existing domestic species.

The possible future economic value of species is hard to predict, since it may be based on products or processes that are as of yet unimaginable. For instance, references to biological diversity as "natural wealth" recently took on new meaning with an amazing report that certain plants can accumulate significant amounts of gold, which may potentially lead to the farming of old mine sites for precious minerals (Anderson et al. 1998). If biological diversity is reduced in the future, the ability of scientists to locate and utilize new species for such purposes likewise will be decreased.

Health agencies and pharmaceutical companies are making a major effort to collect and screen plants and other species for compounds that have the ability to fight human diseases (Eisner and Beiring 1994; Davis 1995; Tuxill 1999). The discovery of a potent anticancer chemical in the Pacific yew (*Taxus brevifolia*), a tree native to North American old-growth forests, is only one recent result of this search. Another species with medicinal value is the ginkgo tree (*Ginkgo biloba*), which occurs in the wild in a few isolated localities in China. During the last 20 years, a $500 million-per-year industry has developed around the cultivation of the ginkgo tree and the manufacture of medicines from its leaves, widely used in Europe and Asia to treat problems of blood circulation, stroke, and memory loss (Figure 1.21; Del Tredici 1991).

The growing biotechnology industry is finding new ways to reduce pollution, develop industrial processes, and fight diseases threatening human health (Frederick and Egan 1994). In some cases, newly discovered or well-known species have been found to have exactly those properties needed to deal with a significant human problem. Innovative techniques of molecular biology are allowing

1.21 Ginkgo trees are cultivated because valuable medicines can be made from their leaves. Each year the woody stems sprout new shoots and branches, which are harvested. This species is the basis of a pharmaceutical business worth hundreds of millions of dollars each year. (Photograph by Peter Del Tredici, Arnold Arboretum.)

unique, valuable genes found in one species to be transferred to another species. Some of the most promising new species being investigated by industrial scientists are the ancient bacteria that live in extreme environments such as deep-sea geothermal vents and hot springs (Jarrell et al. 1999). Bacteria that thrive in unusual chemical and physical environments can often be adapted to special industrial applications of considerable economic value. One of the most important tools of the multibillion-dollar biotechnology industry, the polymerase chain reaction (PCR) technique for multiplying copies of DNA, depends on an enzyme that is stable at high temperatures; this enzyme was originally derived from a bacterium (*Thermus aquaticus*) endemic to natural hot springs in Yellowstone National Park. The companies Hoffman-LaRoche and Perkins-Elmer, owners of the PCR patents, are earning $200 million per year from this technology (Chester 1996).

A question that is currently being actively debated is: Who owns the commercial development rights to the world's biological diversity? In the past, species were freely collected from wherever they occurred; corporations, often in the developed world, then sold the resulting products at a profit. Increasingly, governments in the developing world are demanding that a share of the profits from new products be returned to the countries and local communities where the original specimens were collected (Vogel 1994). Writing treaties and developing procedures to guarantee participation in this process will be a major diplomatic and economic challenge in the coming years.

While most species may have little or no direct economic value, a small proportion may have the potential to supply medical treatments, to support a new industry, or to prevent the collapse of a major agricultural crop. If just one of these species becomes extinct before it is discovered, it will be a tremendous loss to the global economy, even if the majority of the world's species are preserved. Stated another way, the diversity of the world's species can be compared to a manual on how to keep the Earth running effectively. The loss of a species is like tearing a page out of the manual. If we ever need the information from that page to save ourselves and the Earth's other species, we will find that it is irretrievably lost.

Existence value

Many people throughout the world care about wildlife and plants and are concerned with their protection. This concern may be associated with a desire to someday visit the habitat of a unique species and see it in the wild, or it may be a fairly abstract identification. Particular species, so-called "charismatic megafauna" such as pandas, lions, elephants, manatees, bison, and many birds, elicit strong responses in people. People value these emotions in a direct way by joining and contributing money to conservation organizations that work to protect these species and habitats. In the United States, $4 billion was contributed in 1995 to environmental wildlife organizations, with The Nature Conservancy, the World Wildlife Fund, Ducks Unlimited, and the Sierra Club topping the list. Citizens also show their concern by directing their governments to spend money on conservation programs. For example, the government of the United States has already spent $30 million to protect a single rare species, the California condor (*Gymnogyps californianus*).

Such **existence value** can also be attached to biological communities, such as old-growth forests, tropical rain forests, coral reefs, prairies, and coastal marshes, and to areas of scenic beauty. People and organizations contribute large sums of money annually to ensure the continuing existence of these habitats. The money spent to protect biological diversity, particularly in the developed countries of the world, is on the order of billions of dollars per year. This sum represents the existence value of species and biological communities—the amount that people are willing to pay to prevent species from going extinct and habitats from being destroyed.

In summary, ecological economics has helped to draw attention to the wide range of goods and services provided by biological diversity. That has helped scientists to better evaluate projects, because they can now account for important variables—environmental impacts—that were previously left out of the equation. When complete analyses have been done of large-scale development projects, some projects that initially appear to be successful are actually running at an economic loss. For example, to evaluate the success of a development project, such as an irrigation project using water diverted from a tropical wetland ecosystem, the short-term benefits (improved crop yields) must be weighed against the environmental costs. Figure 1.22 shows the total economic value of a tropical wetland ecosystem, including its use value, option value, and existence value. When the wetland ecosystem is damaged by the removal of water, the ecosystem's ability to provide the services shown in the figure are curtailed, their value greatly diminishes, and the economic success of the project is called into question. It is only by incorporating the wetlands' value into this equation that an accurate view of the total project can be gained.

Environmental Ethics

Even though the methods of ecological economics are a positive development for conservation, they can also be viewed as signs of a willingness to accept the present world economic system as it is, with only minor changes. Given a world economic system in which

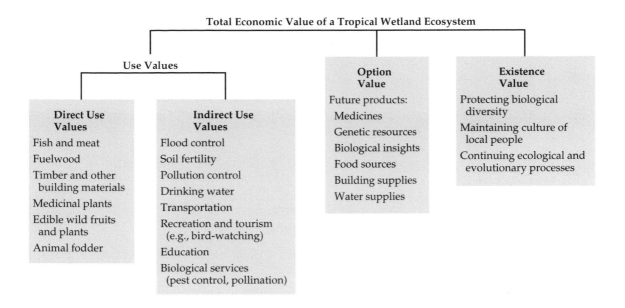

Total Economic Value of a Tropical Wetland Ecosystem

Use Values

Direct Use Values

Fish and meat

Fuelwood

Timber and other building materials

Medicinal plants

Edible wild fruits and plants

Animal fodder

Indirect Use Values

Flood control

Soil fertility

Pollution control

Drinking water

Transportation

Recreation and tourism (e.g., bird-watching)

Education

Biological services (pest control, pollination)

Option Value

Future products:

Medicines

Genetic resources

Biological insights

Food sources

Building supplies

Water supplies

Existence Value

Protecting biological diversity

Maintaining culture of local people

Continuing ecological and evolutionary processes

1.22 Evaluating the success of a project must incorporate the full range of its environmental effects. This figure shows the total economic value of a tropical wetland ecosystem. The value of the wetland ecosystem is reduced when water is removed for irrigation of crops. When that lowered value is taken into account, the irrigation project may represent an economic loss. (Based on data in Barbier 1993.)

millions of children die each year from disease, malnutrition, crime, and war, and in which thousands of unique species become extinct each year due to habitat destruction, we may ask, do we need to make minor adjustments or major structural changes?

A complementary approach to protecting biological diversity and improving the human situation through stringent regulations, incentives, fines, and environmental monitoring is to change the fundamental values of our materialistic society. **Environmental ethics**, a vigorous new discipline within philosophy, articulates the ethical value of the natural world (Van de Veer and Pierce 1994; Armstrong and Botzler 1998). If our society adhered to the principles of environmental ethics, the preservation of the natural environment and maintenance of biological diversity would become fundamental priorities. The natural consequences would be a lowered consumption of resources, greater amounts of land devoted to conservation, and an effort to limit human population growth. Many traditional cultures have successfully coexisted with their environments for thousands of years because of societal ethics that encourage personal responsibility and efficient use of resources, and this could also become a priority for modern societies.

Although economic arguments are often advanced to justify the protection of biological diversity, there are also strong ethical arguments for doing so (Naess 1989; Rolston 1994). These arguments have foundations in the value systems of most religions, philosophies, and cultures, and thus can be readily understood by most people. Ethical arguments for preserving biological diversity appeal to the nobler instincts of people and are based on widely held truths. They may appeal to a general respect for life, a reverence for nature, a sense of the beauty, fragility, uniqueness, or antiquity of the living world, or a belief in divine creation. People will accept or at least consider these arguments on the basis of their belief systems (Callicott 1994).

In contrast, arguments based on economic grounds are still being developed and may eventually prove to be inadequate, highly inaccurate, or unconvincing. Economic arguments by themselves might provide a basis for valuing species, but they might also be used (and misused) to decide that we ought not to save a species, or that we ought to save one species and not another (Bulte and van Kooten 2000). In economic terms, a species that has a small physical size, low population numbers, a limited geographical range, an unattractive appearance, no immediate use to people, and no relationship to any species of economic importance will be given a low value; such qualities may characterize a substantial proportion of the world's species, particularly insects, other invertebrates, fungi, nonflowering plants, bacteria, and protists. Costly attempts to preserve these species may not have any short-term economic justification.

Several ethical arguments, however, can be made for preserving *all species*, regardless of their economic value. The following assertions are important to conservation biology because they provide the rationale for protecting rare species and species of no obvious economic value.

- *Each species has a right to exist.* All species represent unique biological solutions to the problem of survival. On this basis, the survival of each species must be guaranteed, regardless of its abundance or its importance to humans. This is true whether the species is large or small, simple or complex, ancient or recently evolved, economically important or unimportant. All species are part of the community of living beings and have just as much right to exist as humans do. Each species has value for its own sake, an **intrinsic value** unrelated to human needs. Besides not having the right to destroy species, people have the responsibility of taking action to prevent species from going extinct as the result of human activities. This argument envisions humans as moving beyond a limited anthropocentric (human-centered) perspective, becoming part of and identifying with a larger biotic community in which we respect all species and their right to exist.

  How can we assign rights of existence and legal protection to nonhuman species when they lack the self-awareness that is usually associated with the morality of rights and duties? Further, how can nonanimal species, such as mosses and fungi, have rights when they lack even a nervous system to sense their environment? Many advocates for environmental ethics believe that species do assert their will to live through their production of offspring and their continuous evolutionary adaptation to a changing environment. The premature extinction of a species due to human activities destroys this natural process, and can be regarded as a "superkilling" (Rolston 1989) because it kills not only living individuals but also future generations of the species and limits the processes of evolution and speciation.

- *All species are interdependent.* Species interact in complex ways as parts of natural communities. The loss of one species may have far-reaching consequences for other members of the community. Other species may become extinct in response, or the entire community may become destabilized as the result of cascades of species extinction. As we learn more about global processes, we are also finding out that many chemical and physical characteristics of the atmosphere, the climate, and the ocean are linked to biological processes in a self-regulating manner, as set forth in the Gaia hypothesis (Lovelock 1988). If this is the case, our instincts toward self-preservation may impel us to preserve biodiversity. When the natural world prospers, we prosper. We are obligated to conserve the system as a whole because that is the appropriate survival unit.

- *People have a responsibility to act as stewards of the Earth.* Many religious adherents find it wrong to allow the destruction of species, because they are God's creation. If God created the world, then the species God created have value. Within the Jewish, Christian, and Islamic traditions, human responsibility for protecting animal species is explicitly described as part of the covenant with God. Other major religions, including Hinduism and Buddhism, strongly support the preservation of nonhuman life.

- *People have a responsibility to future generations.* From a strictly ethical point of view, if we degrade the natural resources of the Earth and cause species to become extinct, future generations of people will have to pay the price in terms of a lower standard of living and quality of life. Therefore, people of today should use resources in a sustainable manner so as not to damage species and communities. We might imagine that we are borrowing the Earth from future generations, and when they receive it from us they will expect to get it in good condition.

- *Respect for human life and concern for human interests are compatible with a respect for biological diversity.* Some people argue that a concern for preserving nature takes away from a proper concern for human life, but that is false. An appreciation of the complexity of human culture and the natural world leads people to respect and protect all life in its diverse forms. It is also true that people will be more likely to protect biological diversity when they have full political rights, a secure livelihood, and an awareness of environmental issues. Working for the social and political benefit of poor and powerless people is compatible with efforts to preserve the natural environment. Over the long term, human maturity leads naturally to an "identification with all life forms" and "the acknowledgment of the intrinsic value of these forms" (Naess 1986). This view sees an expanding circle of moral obligations, moving outward from oneself to include duties to relatives, one's own social group, all humanity, animals, all species, the ecosystem, and ultimately the whole Earth (Figure 1.23; Noss 1992).

- *Nature has spiritual and aesthetic value that transcends its economic value.* Throughout history, religious thinkers, poets, writers, artists, and musicians of all varieties have drawn inspiration from nature. For many people, an essential quality of this inspiration requires experiencing nature in an undisturbed setting. Simply reading about species or seeing them in museums, gardens, zoos, and nature videos will not suffice. Nearly everyone enjoys wildlife and landscapes aesthetically, and outdoors activities involving nature appreciation are enjoyed by millions of people. A loss in biological diversity diminishes this experience. For instance, if species of whales, wildflowers, and butterflies go

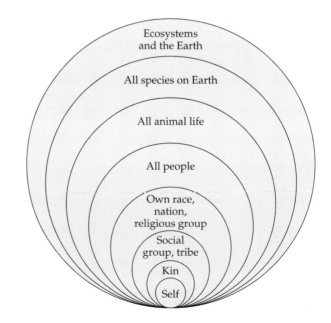

1.23 Environmental ethics holds that an individual has an expanding set of moral obligations, extending outward beyond the self to progressively more inclusive levels. (From Noss 1992.)

extinct in the next few decades, important imagery will be lost to the direct experience of future generations of artists and children.

- *Biological diversity is needed to determine the origin of life.* Three of the central mysteries in the world of science are how life originated, how the diversity of life found on Earth today came about, and how humans evolved. Thousands of biologists are working on these problems and are coming ever closer to the answers. For example, plant taxonomists using molecular and phylogenetic techniques recently determined that a shrub from the Pacific island of New Caledonia represents the only living species in the most ancient lineage of flowering plants (Mathews and Donoghue 1999). However, when such species become extinct, important clues are lost, and the mystery then becomes harder to solve. If humans' closest relatives—chimpanzees, bonobos, gorillas, and orangutans—become extinct in the wild, we will lose important clues in understanding human evolution.

Deep ecology

Throughout the world, environmental activist organizations such as Greenpeace and EarthFirst! are committed to using their knowledge of environmental issues to protect species and ecosystems. One of the most-developed environmental philosophies that supports this activism is described in *Deep Ecology: Living as if Nature Mattered* (Devall and Sessions 1985; Naess 1989; Sessions 1995). **Deep ecology** begins with the premise that all species have value in and of themselves, and that humans have no right to reduce this richness. Because present human activities are destroying the Earth's biological diversity, existing political, economic, technological, and ideological structures must change. These changes entail enhancing the life quality of people, emphasizing improvements in environmental quality, aesthetics, culture, and religion rather than higher levels of material consumption. The philosophy of deep ecology includes an obligation to work to implement the needed changes through political activism and a commitment to personal lifestyle changes. Deep ecology may be the philosophy most compatible with conservation biology because it is not only a set of ideas that can be discussed, it is also a powerful plan for personal, social, and political change.

Summary

1. Conservation biology is a synthesis of scientific disciplines that deals with today's unprecedented biodiversity crisis. It combines basic and applied research approaches to prevent the loss of biodiversity: specifically, the extinction of species, the loss of genetic variation, and the destruction of biological communities.

2. Conservation biology rests on a number of underlying assumptions that are accepted by most conservation biologists: The diversity of species should be preserved; the extinction of species due to human activities should be prevented; the complex interaction of species in natural communities should be maintained; the evolution of new species should be allowed to continue; biological diversity has value in and of itself.

3. The Earth's biological diversity includes the entire range of living species, the genetic variation that occurs among individuals within a species, the biological communities in which species live, and the ecosystem-level interactions of the community with the physical and chemical environment.

4. Certain keystone species are important in determining the ability of other species to persist in a community. Keystone resources, such as water sources and tree holes, may occupy only a small fraction of a habitat, but may be crucial to the persistence of many species in the area.

5. The greatest biological diversity is found in the tropical regions of the world, with large concentrations in tropical rain forests, coral reefs, tropical lakes, and the deep sea. The majority of the world's species have still not been described and named.

6. The new field of ecological economics is developing methods for valuing biological diversity and in the process is providing arguments for its protection. Throughout the world, large development projects, such as dams and industrial facilities, are increasingly being analyzed by means of environmental impact assessments and cost–benefit analyses before being approved.

7. Components of biological diversity can be given direct economic values, which are assigned to products harvested by people, or indirect economic values, which are assigned to benefits provided by biological diversity that do not involve harvesting or destroying the resource. Direct values can be further divided into consumptive use value and productive use value. Consumptive use value is assigned to products that are used locally, such as fuelwood, medicines, wild animals, and building materials; these are often crucial to the livelihoods of people in rural areas.

8. Productive use value can be assigned to products harvested in the wild and sold in regional and national markets, such as commercial timber and fish. Species collected in the wild have great productive use value in their ability to provide new founder stock for domestic species and for the genetic improvement of agricultural crops. Wild species have also been a major source of new medicines.

9. Indirect values can be assigned to aspects of biological diversity that provide economic benefits to people but are not harvested

or damaged during these uses. Nonconsumptive use values of ecosystems include ecosystem productivity, protection of soil and water resources, the positive interactions of wild species with commercial crops, and regulation of climate.

10. Biological diversity is part of the foundation of the outdoor recreation and ecotourism industries. In many countries of the developing world, ecotourism represents one of the major sources of foreign income.

11. Biological diversity also has an option value in terms of its potential to provide future benefits to human society, such as new medicines, industrial products, and crops. It also has an existence value, based on the amount of money people are willing to pay to protect species and habitats.

12. Protecting biological diversity can be justified on ethical grounds as well as on economic grounds. The value systems of most religions, philosophies, and cultures provide justifications for preserving species that are readily understood by people. These justifications support protecting even species that have no obvious economic value to people.

13. The most central ethical argument is that species have a right to exist based on an intrinsic value unrelated to human needs. People do not have the right to destroy species and must take action to prevent the extinction of species.

Suggested Readings

Armstrong, S. and R. Botzler (eds.). 1998. *Environmental Ethics: Divergence and Convergence.* McGraw-Hill, New York. Covers a wide variety of topics that can be used to initiate debate.

Buchmann, S. L. and G. P. Nabhan. 1996. *The Forgotten Pollinators.* Island Press, Washington, D.C. Wild species are vital pollinators of many crop species and endangered plants, as shown in this beautiful book.

Bulte, E. and G. C. van Kooten. 2000. Economic science, endangered species, and biodiversity loss. *Conservation Biology* 14: 113–119. Considers the limitations of economic arguments in preserving biological diversity; good starting point for discussions.

Chapin III, F. S., O. E. Sala, E. C. Buke, et al. 1998. Ecosystem consequences of changing biodiversity. *BioScience* 48: 45–52. Recent investigations throughout the world provide evidence that reduced species diversity lowers ecosystem productivity.

Costanza, R., R. d'Arge, R. de Groot, S. Farber, et al. 1997. The value of the world's ecosystem services and natural capital. *Nature* 387: 253–260. High-profile article by top ecological economists estimates the Earth's total ecosystem services as worth around $32 trillion a year.

Daily, G. C. (ed.). 1997. *Nature's Services: Societal Dependence on Ecosystem Services.* Island Press, Washington, D.C. Clear explanations of why maintaining species and ecosystems is critical to human societies.

Heywood, V. H. 1995. *Global Biodiversity Assessment*. Cambridge University Press, Cambridge. This massive book comprehensively treats the subject, with chapters by leading scientists and a huge bibliography.

Huston, M. A. 1994. *Biological Diversity: The Coexistence of Species on Changing Landscapes*. Cambridge University Press, Cambridge. Extensive review of the patterns and theories of biological diversity.

Kellert, S. R. and E. O. Wilson (eds.). 1993. *The Biophilia Hypothesis*. Island Press, Washington, D.C. Discussion of inherent biological reasons for valuing and cherishing nature.

Meffe, G. C., C. R. Carroll, and contributors. 1997. *Principles of Conservation Biology,* Second Edition. Sinauer Associates, Sunderland, MA. Excellent advanced textbook.

Morell, V. 1999. The variety of life. *National Geographic* 195 (February): 6–32. Special issue on biodiversity includes this and other beautifully illustrated articles about biological diversity, threats to its existence, and key conservation projects.

Plotkin, M. J. 1993. *Tales of a Shaman's Apprentice*. Viking/Penguin, New York. Vivid account of ethnobotanical exploration and efforts to preserve medical knowledge.

Primack, R. 1998. *Essentials of Conservation Biology,* Second Edition. Sinauer Associates, Sunderland, MA. A full-length textbook suitable for undergraduate courses.

Sessions, G. (ed.). 1995. *Deep Ecology for the 21st Century: Readings on the Philosophy and Practice of the New Environmentalism*. Shambala Books, Boston. The key to protecting biological diversity needs to be linked to personal, social, political, and economic changes.

Tilman, D. 1999. The ecological consequences of change in biodiversity: A search for general principles. *Ecology* 80: 1455–1474. Mixture of field data, experiments, and models used to demonstrate relationships among biodiversity, stability, productivity, and susceptibility to invasion, with implications for management.

Tuxill, J. 1999. *Nature's Cornucopia: Our Stake in Plant Diversity*. World Watch Institute, Washington, D.C. Plants are crucial to both traditional and modern societies.

Wilson, E. O. 1992. *The Diversity of Life*. Belknap Press of Harvard University Press, Cambridge, MA. An outstanding description of biological diversity, written for the general public.

Wilson, E. O. and D. L. Perlmann. 1999. *Conserving Earth's Biodiversity*. Island Press, Washington, D.C. CD-ROM with video clips, maps, models, and essays.

World Conservation Monitoring Centre (WCMC). 1992. *Global Biodiversity: Status of the Earth's Living Resources*. Compiled by the World Conservation Monitoring Centre. Chapman and Hall, London. Huge sourcebook of biodiversity facts and figures. Strong section on environmental economics.

Chapter *2*

Threats to Biological Diversity

A healthy environment has great economic, aesthetic, and ethical value. Maintaining a healthy environment means preserving all of its components in good condition: ecosystems, communities, species, and genetic variation. For each of these components, initial threats can eventually lead to complete loss. Communities can be degraded and reduced in area and much of their ecosystem value lost, and eventually they can even be completely destroyed. But as long as all of their original species survive, communities still have the potential to recover. Similarly, genetic variation within a species will be reduced as population size is lowered, and this can lead to genetic problems from which the species cannot recover. But species can potentially regain genetic variation through mutation, natural selection, and recombination, following a successful rescue program. However, when a species goes extinct, the unique genetic information contained in its DNA and the special combination of characters that it possesses are forever lost. Once extinction occurs, a species' populations cannot be restored, the communities to which it once belonged are impoverished, and the species' potential value to humans can never be realized.

Rates of Extinction

The term "extinct" has many nuances, and its meaning can vary somewhat depending on the context. A species is considered *extinct* when no member of the species remains alive anywhere in the world: "The Bachman's warbler is extinct" (Figure 2.1). If individuals of a species remain alive only in captivity or in other human-controlled situations, the species is said to be *extinct in the wild*: "The Franklin tree is extinct in the wild but grows well under cultivation" (see Figure 2.1). In both of these situations, the species would

2.1 One of the first Neotropical migrant songbirds to become extinct as a result of tropical deforestation was the Bachman's warbler (*Vermivora bachmanii*), last seen in the 1960s. The Cuban forests in which this species overwintered were almost entirely cleared for sugarcane fields. The warbler is shown in this Audubon print with the flowering Franklin tree (*Franklinia altamaha*), which is now extinct in the wild, although it can still be found in arboretums and other cultivated gardens. (By John James Audubon; photograph from the Ewell Sale Stewart Library, The Academy of Natural Sciences of Philadelphia.)

be considered *globally extinct*. A species is considered *locally extinct* when it is no longer found in an area that it once inhabited but is still found elsewhere in the wild: "The American burying beetle once occurred throughout eastern and central North America; it is now locally extinct in all but three widely separated areas" (Figure 2.2). Some conservation biologists speak of a species being *ecologically extinct* if it persists at such reduced numbers that its effects on the other species in its community are negligible: "The tiger is ecologically extinct: so few tigers remain in the wild that their effect on prey populations is insignificant."

A vital question in conservation biology is, How long will it take for a given species to become extinct following a severe reduction in its range or the degradation or fragmentation of its habitat? When populations fall below a certain critical number of individuals, they are very likely to become extinct. In some populations, a few individuals might persist for years or decades, and even reproduce, but their ultimate fate would be extinction unless strong conservation measures were taken to promote their recovery. In woody plants in particular, isolated, nonreproductive individuals can persist for hundreds of years. These species have been called "the living dead": even though technically the species is not extinct while a few

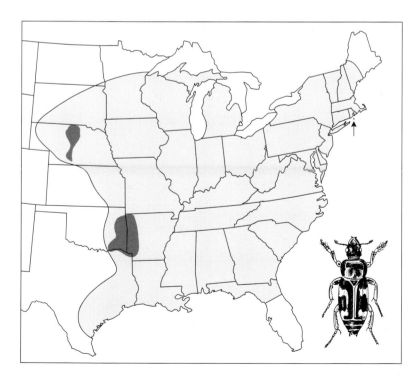

2.2 The American burying beetle (*Nicrophorus americanus*) was once widespread in the eastern and central United States (shaded area) but is now found in only three isolated populations (black areas; Block Island, in Long Island Sound, is highlighted with an arrow). Intensive efforts have been initiated to determine the cause of this decline and develop a recovery plan. (After Kozol et al. 1994.)

individuals live, the population is no longer reproductively viable, so the species' future is limited to the life spans of the remaining individuals (Janzen 1986b). In order to successfully preserve species, conservation biologists must identify the human activities that affect the stability of populations and drive species to extinction. They must also determine the factors that make populations vulnerable to extinction.

Human-Caused Extinctions

The global diversity of species reached an all-time high in the present geological period. The most advanced groups of organisms—insects, vertebrates, and flowering plants—reached their greatest diversity about 30,000 years ago. However, since that time species richness has decreased as human populations have grown (Leaky and Lewin 1996). At present, a phenomenal 40% of the total net primary productivity (the living material produced by plants) of the terrestrial environment is used or wasted in some way by people; this represents about 25% of the total primary productivity of the Earth. People are also playing an increasingly dominant role in other ecosystem components, such as the nitrogen cycle and carbon dioxide levels (Table 2.1).

The first noticeable effects of human activity on extinction rates can be seen in the elimination of large mammals from Australia and North and South America at the time humans first colonized these continents many thousands of years ago. Shortly after humans arrived, 74% to 86% of the megafauna—mammals weighing more

TABLE 2.1 Three ways in which humans dominate the global ecosystem

1. Land Surface
Human land use and demand for resources have transformed as much as half of the Earth's ice-free land surface.

2. Nitrogen Cycle
Each year human activities, such as cultivating nitrogen-fixing crops, using nitrogen fertilizers, and burning fossil fuels, release more nitrogen into terrestrial systems than is added by natural biological and physical processes.

3. Atmospheric Carbon Cycle
By the middle of the twenty-first century, human use of fossil fuels will have resulted in a doubling of the level of carbon dioxide in the Earth's atmosphere.

Source: Data from Vitousek 1994; Vitousek et al. 1997.

than 44 kg (100 lbs.)—in these areas became extinct. These extinctions were probably caused directly by hunting (Martin and Klein 1984; Miller et al. 1999) and indirectly by burning, clearing forests, and the spread of introduced diseases. On all continents and on numerous islands, there is an extensive record of prehistoric human alteration and destruction of habitat that coincides with a high rate of species extinctions.

How has human activity affected extinction rates in more recent times? Extinction rates are best known for birds and mammals because these species are relatively large, well studied, and conspicuous. Extinction rates for the other 99.9% of the world's species are just rough guesses at the present time. However, extinction rates are uncertain even for birds and mammals, because some species that were considered extinct have been rediscovered, and other species that are presumed to be **extant** (still living) may actually be extinct. Based on the available evidence, the best estimate is that about 85 species of mammals and 113 species of birds have become extinct since the year 1600 (Table 2.2), representing 2.1% of mammal species and 1.3% of birds (Smith et al. 1993; Heywood 1995). While these numbers initially may not seem alarming, the trend of these extinction rates is upward, with a dramatic increase in extinctions having occurred in the last 150 years (Figure 2.3). The extinction rate for birds and mammals was about one species every decade during the period from 1600 to 1700, but it rose to one species every

TABLE 2.2   Recorded extinctions, 1600 to the present

Taxon	Recorded extinctions[a]				Approximate number of species	Percentage of taxon extinct
	Mainland[b]	Island[b]	Ocean	Total		
Mammals	30	51	4	85	4000	2.1
Birds	21	92	0	113	9000	1.3
Reptiles	1	20	0	21	6300	0.3
Amphibians[c]	2	0	0	2	4200	0.05
Fishes[d]	22	1	0	23	19,100	0.1
Invertebrates[d]	49	48	1	98	1,000,000+	0.01
Flowering plants[e]	245	139	0	384	250,000	0.2

Source: After Reid and Miller 1989; data from various sources.

[a]Numerous additional species have presumably gone extinct without ever being recorded by scientists.

[b]Mainland areas are those with landmasses of 1 million km[2] or greater (the size of Greenland or larger); smaller landmasses are considered islands.

[c]There has been an alarming decrease in amphibian populations in the last 20 years; some scientists believe that many amphibian species are on the verge of becoming extinct or already extinct.

[d]The figures given are primarily representative of North America and Hawaii.

[e]The numbers for flowering plants include extinctions of subspecies and varieties as well as species.

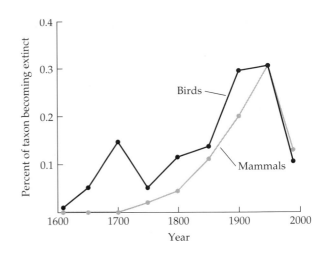

2.3 Extinction rates during 50-year intervals for birds and mammals since 1600. Extinctions have increased from 1800 to 1950, with some evidence of a decline during the last 50 years. (Data from Smith et al. 1993.)

year during the period from 1850 to 1950. This increase in the rate of species extinction is an indication of the seriousness of the threat to biological diversity.

Some evidence suggests a decline in the extinction rate for birds and mammals during the past few decades (see Figure 2.3). This may be due in part to recent efforts to save species from extinction, but it can also be attributed to a procedure adopted by international organizations to list a species as extinct only when it has not been seen for 50 years, or when exhaustive surveys have failed to locate any remaining individuals. Many species not yet technically extinct have been decimated by human activities and persist only in very low numbers. These species may be considered ecologically extinct in that they no longer play a role in community organization. The future of many such species is doubtful.

About 11% of the world's remaining bird species are threatened with extinction; similar percentages hold for mammal and tree species. Table 2.3 shows certain animal groups for which the danger is even more severe, such as the family of lizards known as iguanas (Mace 1994). The threat to some freshwater fishes and mollusks may be equally severe (Williams and Nowak 1993). Plant species are also threatened, with gymnosperms (conifers, ginkgos, and cycads) and palms among the especially vulnerable groups. While extinction does occur as a natural process, more than 99% of modern species extinctions can be attributed to human activity (Lawton and May 1995).

Extinction rates in water and on land

Of all the animal and plant species known to have gone extinct in the period from 1600 to the present, almost half were island species

TABLE 2.3 Numbers of species threatened with extinction in major groups of animals and plants, and some key families and orders

Group	Approximate number of species	Number of species threatened	Percentage of species threatened
Vertebrate Animals			
Fishes	24,000	452	2
Amphibians	3000	59	2
Reptiles	6000	167	3
Boidae (constrictor snakes)	17[a]	9	53
Varanidae (monitor lizards)	29[a]	11	38
Iguanidae (iguanas)	25[a]	17	68
Birds	9500	1029	11
Anseriformes (waterfowl)	109[a]	36	33
Psittaciformes (parrots)	302[a]	118	39
Mammals	4500	505	11
Marsupialia (marsupials)	179[a]	86	48
Canidae (wolves and wild dogs)	34[a]	13	38
Cervidae (deer)	14[a]	11	79
Plants			
Gymnosperms	758	242	32
Angiosperms (flowering plants)	240,000	21,895	9
Palmae (palms)	2820	925	33

Source: Data from Smith et al. 1993 and Mace 1994.

[a]Number of species for which information is available.

(see Table 2.2), even though islands represent only a small fraction of the Earth's surface. In contrast, it has been estimated that only nine species—three species of marine mammals, five marine birds, and four mollusk species—have become extinct in the world's vast oceans during modern times (Carlton et al. 1999). Further, there are no documented cases of fish or coral species having gone extinct during this period. This number of extinctions in the marine environment is almost certainly an underestimate, because marine species are not as well known as terrestrial species, but it may also reflect the greater resiliency of marine species in response to disturbance. However, the significance of these marine losses may be greater than the numbers suggest because many marine mammals are top predators that have a major effect on marine communities. Also, some marine species are the sole representatives of their genus, family, or even order, so the extinction of even a few marine species can represent a serious loss to global biological diversity.

The majority of freshwater fish and flowering plant extinctions have occurred in mainland areas simply because of the vastly greater numbers of species in these regions. In a survey of the rich freshwater fish fauna of the Malay Peninsula, only 122 of the 266 species of fish known to exist on the basis of earlier collections could still be found (Mohsin and Ambak 1983). In North America, over one-third of the freshwater fish species are in danger of extinction (Moyle and Leidy 1992; Moyle 1995). The danger to aquatic invertebrates, such as freshwater mussels and crayfish, is particularly high due to the impact of dams, pollution, introduced species, and habitat destruction.

Extinction rates on islands

The highest species extinction rates during historic times have occurred on islands (see Table 2.2). Most of the known extinctions of birds, mammals, and reptiles during the last 350 years have occurred on islands (WCMC 1992; Pimm et al. 1995), and more than 80% of the endemic plants of some oceanic islands are extinct or in danger of extinction. A species is **endemic** to the location where it occurs naturally. Island species are particularly vulnerable to extinction because many of them are endemic to only one or a few islands and have only one or a few local populations. Recorded extinction rates may also be higher on islands simply because these areas are better studied than continental areas.

A species may be endemic to a wide geographical area, like the black cherry tree (*Prunus serotina*), which is found across North America, Central America, and South America; or a species may be endemic to a small geographical area, like the giant Komodo dragon (*Varanus komodoensis*), which is known only from several small islands in the Indonesian archipelago. A more extreme example is the Mauna Kea silversword (*Argyroxiphium sandwicense sandwicense*), a plant found in nature in only one volcanic crater on the island of Hawaii. Isolated geographical units, such as remote islands, old lakes, and solitary mountain peaks, often have high percentages of endemic species. In contrast, geographical units of equivalent area that are not isolated typically have much lower percentages of endemic species. One of the most notable examples of an isolated area with a high rate of endemism is the island of Madagascar (Myers 1986). Here, the tropical moist forests are spectacularly rich in endemic species: 93% of the 28 primate species, 99% of the 144 species of frogs, and over 70% of the plant species on the island are found nowhere else. Even higher percentages of endemic plant species are found in New Zealand. By comparison, around 1% of the plant species that occur in the United Kingdom and the Solomon Islands are endemics (Table 2.4). If the communities on Madagascar or other isolated islands are destroyed or damaged, or populations are intensively harvested, then these endemic species will become extinct. In contrast, mainland species often have many

TABLE 2.4 Number of plant species and their status for various islands and island groups

Island(s)	Native species	Endemic species	Percentage endemic	Number of species threatened	Percentage of species threatened
Solomon Islands	2780	30	1	43	2
United Kingdom	1500	16	1	28	2
Sri Lanka	3000	890	30	436	15
Jamaica	2746	923	33	371	14
Philippines	8000	3500	44	371	5
Cuba	6004	3229	54	811	14
Fiji	1307	760	58	72	6
Madagascar	9000	6500	72	189	2
New Zealand	2160	1942	90	236	11
Australia	15,000	14,074	94	1597	11

Source: WRI 1998.

populations over a wide area, so that the loss of one population is not catastrophic for the species. Even in mainland areas, however, certain local regions are noted for their concentrations of endemic species, resulting from such factors as geological age and a wide variety of habitats (Figure 2.4).

One intensively studied island group is the Hawaiian archipelago (Olson 1989; Pimm et al. 1995). There were 98 species of endemic birds in the Hawaiian islands before the arrival of the Polynesians in 400 A.D. The Polynesians introduced the Polynesian rat, the domestic dog, and the domestic pig; they also began clearing the forest for agriculture. As a result of increased predation and dis-

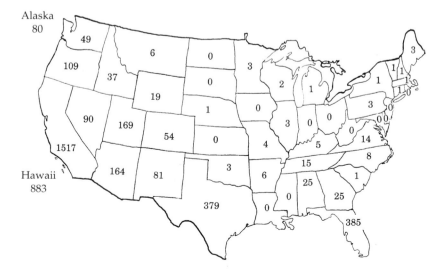

2.4 The number of plant species endemic to the different states in the United States varies greatly. For example, 379 plant species are found only in Texas and nowhere else in the United States. New York, in contrast, has only one endemic species. California, with its large area and vast array of habitats, including deserts, mountains, seacoasts, old-growth forests, and myriad others, is home to more endemic species than any other state. The island archipelago of Hawaii, far from the mainland, hosts many endemic species despite its small area. (From Gentry 1986.)

turbance, more than half of the bird species became extinct prior to the arrival of Europeans in 1778. The Europeans brought cats, new species of rats, the Indian mongoose, goats, cattle, and the barn owl. They also unwittingly brought bird diseases, and they cleared even more land for agriculture and human settlements. Since then, many additional bird species have become extinct and other species are near extinction. Plant species in the Hawaiian islands and on many other islands are also threatened with extinction, mainly through habitat destruction. In Hawaii, fully 91% of the naturally occurring plant species are endemic to the islands. About 10% of these endemic species have become extinct, and 40% of the remaining endemics are threatened with extinction.

It is important to note that whereas past extinctions have occurred predominantly on islands, future extinctions will increasingly take place in continental areas as tropical forests are cleared for human activities (Manne et al. 1999).

Island Biogeography and Modern Extinction Rates

Studies of island communities have led to the development of general rules on the distribution of biological diversity, synthesized as the **island biogeography model** of MacArthur and Wilson (1967). The central observation that this model seeks to explain is the **species–area relationship**: Islands with large areas have more species than islands with small areas (Figure 2.5). This relationship makes intuitive sense, because large islands will tend to have a greater variety of local environments and community types than small islands. Also, larger islands allow greater geographical isolation and a larger number of populations per species, increasing the likelihood of speciation and decreasing the probability of extinction of newly evolved as well as recently arrived species.

The island biogeography model has been used to predict the number and percentage of species that would become extinct if habitats were destroyed (Simberloff 1992; Quammen 1996). It is assumed that if an island has a certain number of species, reducing the area of natural habitat on the island would result in the island being able to support only a number of species corresponding to that on a smaller island (Figure 2.6). This model has been extended from islands to apply to national parks and nature reserves that are surrounded by damaged habitat. These reserves can be viewed as **habitat islands** in an inhospitable "sea" of unsuitable habitat. The model predicts that when 50% of an island (or a habitat island) is destroyed, approximately 10% of the species occurring on the island will be eliminated. If these species are endemic to the area, they will become extinct. When 90% of the habitat is destroyed, 50% of the species will be lost; and when 99% of the habitat is gone,

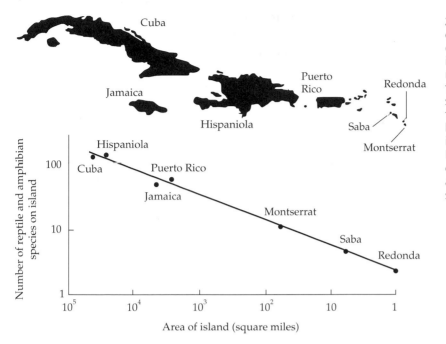

2.5 The number of species on an island can be predicted from the area of the island. In this figure, the number of species of reptiles and amphibians is shown for seven islands in the West Indies. The number of species on large islands such as Cuba and Hispaniola far exceeds that on the tiny islands of Saba and Redonda. (From Wilson 1989.)

about 75% of the original species will be lost. Using these models for a specific example, it has been predicted that 30% of the forest primates in African countries will eventually go extinct because of habitat loss (Cowlishaw 1999). Comparing predictions of species loss with historical examples from forests in Kenya, estimates have

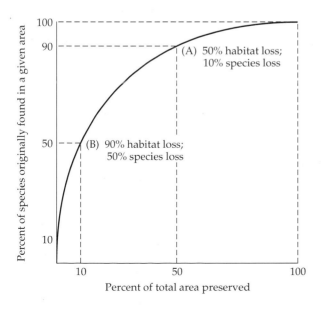

2.6 According to the island biogeography model, the number of species present in an area increases asymptotically to a maximum value. This means that if the area of habitat is reduced by 50%, the number of species lost may be 10% (A); if the habitat is reduced by 90%, the number of species lost may be 50% (B). The shape of the curve is different for each region of the world and for each group of species, but this model gives a general indication of the effect of habitat destruction on species extinctions and the persistence of species in the remaining habitat.

been made of the rates at which remaining forest fragments will lose their bird species. Of the species that will eventually be lost from the fragments, the best estimates predict that half will be lost in 50 years from a 1000 ha fragment, while half will be lost in 100 years from a 10,000 ha fragment (Brooks et al. 1999).

Predictions of extinction rates based on habitat loss vary considerably because each group of species and each geographical area forms characteristic species–area relationships. Because tropical forests account for the great majority of the world's species, estimating present and future rates of species extinction in rain forests gives an approximation of global rates of extinction. If deforestation continues until all of the tropical forests except those in national parks and other protected areas are cut down, about two-thirds of all plant and bird species will be driven to extinction (Simberloff 1986).

Using a conservative value of 1% of the world's rain forests being destroyed per year, Wilson (1989) estimated that 0.2%–0.3% of all species—20,000 to 30,000 species if based on a total of 10 million species—would be lost per year. In more immediate units, 68 species would be lost each day, or 3 species each hour. Over the ten-year period from 2000 to 2010, approximately 250,000 species would become extinct. Other methods using the rates of extinction in tropical rain forests estimate a loss of between 2% and 11% of the world's species per decade (Reid and Miller 1989; Koopowitz et al. 1994). Intensive studies of particular groups of terrestrial vertebrates also yield predictions of alarming extinction rates over the coming decades (Mace 1995). The variation in estimated extinction rates is caused by different estimates of the rate of deforestation, different values for the species–area relationships, and different mathematical approaches (Heywood et al. 1994). Extinction rates might be somewhat lower if species-rich areas could be targeted for conservation efforts; they might also be higher because the highest rates of deforestation are occurring in countries with large concentrations of rare species (Balmford and Long 1994). Regardless of which figure is the most accurate, all of these estimates indicate that hundreds of thousands of species are headed for extinction within the next 50 years.

In addition to the global extinctions that are the primary focus of conservation biology, many species are experiencing a series of local extinctions across their individual ranges. Formerly widespread species are now often restricted to a few small pockets of their former habitats. For example, the American burying beetle (*Nicrophorus americanus*), which was once found all across central and eastern North America, is now found only in three isolated populations (see Figure 2.2). Biological communities are impoverished by such local extinctions. For example, the Middlesex Fells, a local conservation area in metropolitan Boston, contained 338 native plant species in 1894; only 227 native species remained when the area was surveyed 98 years later (Drayton and Primack 1996). Fourteen of the plant species that were lost had been listed as "common" in 1894. An analysis of butterflies in one county in Britain showed that local extinctions over the last 25 years had eliminated 67% of the previously known populations, an astonishingly high rate of local extinction (Thomas and Abery 1995). These large numbers of local extinctions serve as important biological warning signs that something is wrong with the environment. Action is needed to prevent further local extinctions as well as the global extinctions of species that are known to be occurring on a massive and widespread scale.

Causes of Extinction

If species and communities are adapted to local environmental conditions, why are they facing extinction? Shouldn't species and communities tend to persist in the same place over time? The answers to these questions are obvious: Massive disturbances caused by people have altered, degraded, and destroyed the landscape on a vast scale, driving species and even whole communities to the point of extinction. The major threats to biological diversity that result from human activity are habitat destruction, habitat fragmentation, habitat degradation (including pollution), global climate change, the overexploitation of species for human use, the invasion of exotic species, and the increased spread of disease. Most threatened species face at least two or more of these problems, which speed the way to extinction and hinder efforts to protect them (Wilcove et al. 1998; Terborgh 1999; Stearns and Stearns 1999).

These seven threats to biological diversity are all caused by an ever-increasing use of the world's natural resources by the exponentially expanding human population. Until the last few hundred years, the rate of human population growth was relatively slow, with the birth rate only slightly exceeding the mortality rate. The greatest destruction of biological communities has occurred during the last 150 years, during which the human population grew from 1 billion in 1850, to 2 billion in 1930, and reached 6 billion on October 12, 1998 (Figure 2.7). It will climb to an estimated 10 billion by the year 2050. Human numbers have increased because birth rates have remained high while mortality rates have declined as a result of both modern medical discoveries (specifically, the control of dis-

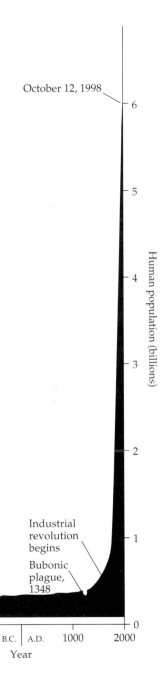

2.7 The human population has increased spectacularly since the seventeenth century. At current growth rates, the population will double in less than 40 years.

ease) and the presence of more reliable food supplies. Population growth has slowed in the industrialized countries of the world but is still high in many areas of tropical Africa, Latin America, and Asia—areas where the greatest biological diversity is also found.

Human population growth by itself is partially responsible for the loss of biological diversity (Krebs et al. 1999). People use natural resources such as fuelwood, wild meat, and wild plants, and they convert vast amounts of natural habitat to agricultural and residential purposes. Some scientists have argued that controlling the size of the human population is the key to protecting biological diversity (Hardin 1993; Meffe et al. 1993). However, human population growth is not the only cause of species extinction and habitat destruction. Examination of the situation in both developing and industrial nations reveals that species extinctions and the destruction of ecosystems are not always caused by individual citizens pursuing their basic needs. The rise of industrial capitalism and materialistic modern societies has produced an accelerated demand for natural resources, particularly in the developed countries. Inefficient and unequal usage of natural resources is also a major cause of the decline in biological diversity. In many countries there is extreme inequality in the distribution of wealth, with the majority of the wealth (money, good farmland, timber resources, etc.) owned and used by a small percentage of the population. As a result, rural people are forced to destroy biological communities and hunt endangered species to extinction because they are poor and have no land or resources of their own (Skole et al. 1994).

In many cases, the causes of habitat destruction are the large-scale industrial and commercial activities associated with a global economy—such as mining, cattle ranching, commercial fishing, forestry, plantation agriculture, manufacturing, and dam construction—that are initiated with the goal of making a profit (Myers 1996). Many of these projects are sanctioned, encouraged, and even subsidized by national governments and international development banks and are touted as providing jobs, commodities, and tax revenues. However, their use of natural resources is often neither efficient nor cost-effective because the emphasis in these industries is on short-term gains; such gains are often made at the expense of the long-term sustainability of the natural resources involved, and generally reflect little regard for the local people who depend on the resources.

The responsibility for the destruction of biological diversity in species-rich tropical areas also lies in the unequal use of natural resources worldwide. People in industrialized countries (and the wealthy minority in the developing countries) consume a disproportionate share of the world's energy, minerals, wood products, and food. Each year, the average citizen of the United States uses 43 times more petroleum products, 34 times more aluminum, and 386 times more paper products than the average citizen of India (WRI 1994). This excessive consumption of resources is not sustainable in the long run, and if this pattern is adopted by the expanding middle class in the developing world, it will cause massive environmental disruption. The affluent citizens of developed countries need to confront their excessive consumption of resources and reevaluate their own lifestyles as they offer aid to curb population growth and protect biological diversity in the developing world (Figure 2.8).

Habitat destruction

The major threat to biological diversity is loss of habitat, and the most important means of protecting biological diversity is habitat preservation. Habitat loss includes habitat destruction as well as habitat damage associated with pollution and habitat fragmentation. Habitat loss is known to be the primary threat to the majority of plants and animals currently facing extinction, with the negative effects of alien species and overexploitation being other important factors (Table 2.5). In many parts of the world, particularly on islands and where human population density is high, most of the original habitat has been destroyed. More than 50% of the primary

2.8 Citizens of wealthy developed countries often criticize the poorer developing nations for a lack of sound environmental policies, but seem unwilling to acknowledge that their own excessive consumption of resources is a major part of the problem. (Cartoon by Scott Willis, © *San Jose Mercury News.*)

TABLE 2.5 Factors responsible for putting threatened species of the United States at risk of extinction

Threatened species group	Percentage of species affected by each factor[a]				
	Habitat degradation and loss	Pollution	Over-exploitation	Competition/predation from alien species	Disease
All species (1880 species)	85	24	17	49	3
All vertebrates (494 species)	92	46	27	47	8
Mammals (85 species)	89	19	47	27	8
Birds (98 species)	90	22	33	69	37
Amphibians (60 species)	87	47	17	27	0
Fishes (213 species)	97	90	15	17	0
All invertebrates (331 species)	87	45	23	27	0
Freshwater mussels (102 species)	97	90	15	17	0
Butterflies (33 species)	97	24	30	36	0
Plants (1055 species)	81	7	10	57	1

Source: Data from Wilcove et al. 1998.

[a]Species may be affected by more than one factor; therefore, rows do not sum to 100%. For example, 87% of threatened amphibian species are affected by habitat degradation and loss, and 47% of these same species are also affected by pollution.

forest wildlife habitat has been destroyed in many Old World countries that are key for biological diversity, such as Kenya, Madagascar, India, the Philippines, and Thailand (Table 2.6). The biologically rich countries of the Democratic Republic of the Congo (formerly Zaire) and Zimbabwe are relatively better off, still having more than half of their wildlife habitat.

Present rates of deforestation vary considerably among countries, with particularly high annual rates of over 2% reported in such tropical countries as Malaysia (2.4%), the Philippines (3.5%), Thailand (2.6%), Costa Rica (3.1%), El Salvador (3.3%), Haiti (3.5%), Honduras (2.3%), Nicaragua (2.5%), Panama (2.2%), and Paraguay (2.6%) (WRI 1998). As a result of habitat fragmentation, farming, logging, and other human activities, very little frontier forest—intact blocks of undisturbed forest large enough to support all aspects of biodiversity—remains in most Old World tropical countries. In the New World, the situation is somewhat better; 42% of Brazilian forest and 59% of Venezuelan forest are frontier forest. In the Mediterranean region, which has been densely populated for thousands of years, only 10% of the original forest cover remains.

For many important wildlife species, the majority of the habitat in their original ranges has been destroyed, and very little of the remaining habitat is protected. For example, the orangutan (*Pongo*

TABLE 2.6 Loss of forest habitat in some countries of the Old World tropics

Country	Current forest remaining (× 1000 ha)	Percentage of habitat lost	Percentage of current forest as frontier forest
Africa			
Democratic Republic of the Congo	135,071	40	16
Gambia	188	38	0
Ghana	1694	91	0
Kenya	3423	82	0
Madagascar	6940	87	0
Rwanda	291	84	0
Zimbabwe	15,397	33	0
Asia			
Bangladesh	862	92	4
India	44,450	80	1
Indonesia	88,744	35	28
Malaysia	13,007	36	14
Myanmar (Burma)	20,661	59	0
Philippines	2402	94	0
Sri Lanka	1581	82	12
Thailand	16,237	78	5
Vietnam	4218	83	2

Source: WRI 1998.

pygmaeus), a great ape that lives in Sumatra and Borneo, has lost 63% of its habitat and is protected in only 2% of its original range.

Threatened rain forests. The destruction of tropical rain forests has come to be synonymous with the loss of species. Tropical moist forests occupy 7% of the Earth's total land surface, but are estimated to contain over 50% of its species. The original extent of tropical rain forests and related moist forests has been estimated at 16 million km^2, based on current patterns of rainfall and temperature (Figure 2.9; Myers 1991b; Sayer and Whitmore 1991; WRI 1994). A combination of ground surveys, aerial photos, and remote sensing data from satellites showed that in 1982 only 9.5 million km^2 remained, an area about equal in size to the continental United States. Another census in 1991 showed a loss of another 2.8 million km^2 during this nine-year period. At the present time, as much as 140,000 km^2 of rain forest is being lost per year—an area larger than the state of Tennessee or the country of Guatemala—with about half being completely destroyed and the remainder degraded to the point at which species composition and ecosystem processes are greatly altered. Unfortu-

2.9 Tropical rain forests are found predominantly in wet, equatorial regions of America, Africa, and Asia. Eight thousand years ago, tropical forests covered the entire shaded area, but human activities have resulted in the loss of a great deal of forest cover, shown in the darkest shade. In the lighter shaded area forests remain, but they are no longer tropical forests; instead, they are (1) secondary forests that have grown back following cutting, (2) plantation forests such as rubber and teak, or (3) forests degraded by logging and fuelwood collection. Only in the regions shown in black are there still blocks of intact natural tropical forest large enough to support all of their biodiversity. (After Bryant et al. 1997.)

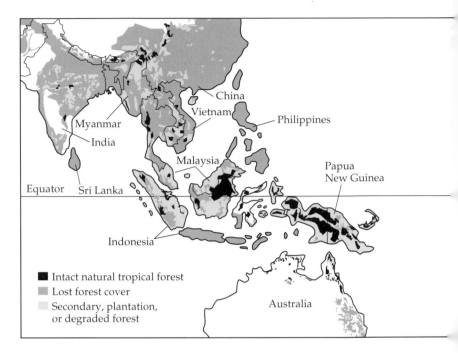

nately, tropical forest ecosystems are easily degraded because their soils are often thin and nutrient-poor, and they are readily eroded by heavy rainfall.

At present, there is considerable discussion in the scientific literature about the original extent and current area of tropical forests as well as the rates of deforestation. Despite the difficulty of obtaining accurate numbers, there is a general consensus that tropical deforestation rates are alarmingly high. Extending the projections forward reveals that at the current rate of loss, there will be very little intact tropical forest left after the year 2040 except in the relatively small areas under protection. The situation is actually even more grim than these projections indicate because the human population is still increasing and poverty is on the rise, putting ever greater demands on the dwindling supply of rain forest.

On a global scale, about 61% of rain forest destruction results from small-scale cultivation of crops by poor farmers. Some of this land is converted into permanent farmland and pastures, but much of it is used for **shifting cultivation** (also known as swidden agriculture or slash-and-burn agriculture), in which patches of forest are cleared, burned, and cultivated for a few seasons, until the soil fertility declines to the point at which the land must be abandoned. The land then returns to secondary forest. Included in this category of land

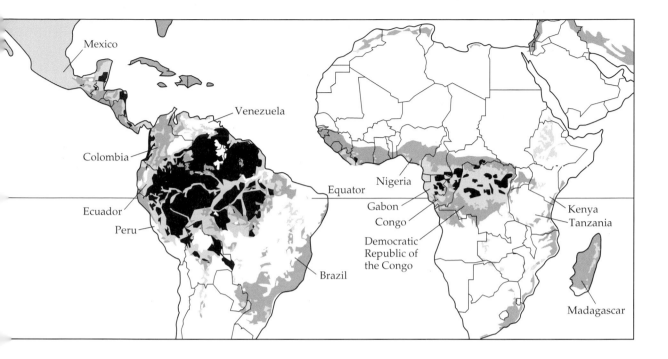

converted to shifting cultivation are forests degraded for fuelwood production, mostly to supply local villagers nearby with wood for cooking fires. More than 2 billion people cook their food with firewood, so their impact is significant. Commercial clear-cutting and selective logging operations cause another 21% of annual loss of tropical forests. Clearing for cattle ranches accounts for 11% of forest loss, with clearing for cash-crop plantations (oil palm, cocoa, rubber, teak, etc.) plus road building, mining, and other activities accounting for the remaining 7% (Figure 2.10). The relative importance of these activities varies by geographical region, with logging being a more significant activity in tropical Asia and America, cattle ranching being most prominent in tropical America, and farming more important for the rapidly expanding human populations in tropical Africa and Asia (Bawa and Dayanandan 1997; Nepstad et al. 1999).

Logging, farming, and cattle ranching on rain forest lands are often related to demand in industrialized countries for cheap agricultural and forestry products, such as rubber, palm oil, cocoa, timber, plywood, and beef. One famous example is the "hamburger connection," in which a large demand for inexpensive beef in the United States and other developed countries in the 1980's provided the economic incentive for clearing large areas of Central and South America for cattle ranches. Consumer boycotts of American fast

2.10 Tropical forests in the Brazilian Amazon are being cut down to make way for cattle ranches. This photograph shows massive deforestation, the result of a "land rush" in Rondonia that occurred when the government provided developers with lucrative tax subsidies and access to the land. Similar scenes are common in Southeast Asia, where logging by timber companies is followed by shifting cultivation and plantation agriculture. (Photograph © Mauricio Simonetti, D. Donne Bryant Stock Photography.)

food chains helped to stop beef purchases from these countries, but had little apparent effect on deforestation, since the beef was sold elsewhere.

A number of places can be singled out as examples of how rapid and serious rain forest destruction can be:

• *Madagascar*. The moist forests of Madagascar, with their rich array of endemic species, particularly 28 lemur species, originally covered 112,000 km². By 1985, they had been reduced to 38,000 km² by a combination of shifting cultivation, cattle grazing, and fire. The present rate of deforestation on Madagascar is about 1100 km² per year, which means that by the year 2020 there may be no moist forests left except in the 1.5% of the island under protection (Green and Sussman 1990). Because Madagascar is the only place where lemurs occur in the wild, the loss of Madagascar's forests will result in the extinction of many lemur species (Figure 2.11).

2.11 The aye-aye (*Daubentonia madagascariensis*) is one of the endangered endemic lemur species of Madagascar. The aye-aye is the subject of conservation efforts in the field as well as captive breeding programs. All 28 lemur species—an entire order of primates found nowhere else in the world—are in danger of becoming extinct on Madagascar. (Photograph © David Haring, Duke University Primate Research Center.)

- *The Atlantic Coast of Brazil.* Another area of high endemism is the Atlantic Coast forest of eastern Brazil. Fully half of its tree species are endemic to the area, and the region supports a number of rare and endangered animals, including the golden lion tamarin (*Leontopithecus rosalia*). In recent decades, the Atlantic Coast forest has been almost entirely cleared for sugar- cane, coffee, and cocoa production; less than 9% of the original forest remains (Brooks and Balmford 1996). The remaining for- est is not in one large block, but is divided into isolated frag- ments that may be unable to support viable populations of many wide-ranging species. The single largest patch of forest remaining is only 7000 km^2, and this patch is highly disturbed in places.

- *Coastal Ecuador.* The coastal region of Ecuador originally was covered by a rich forest filled with endemic species. It was mini- mally disturbed by human activity until 1960. At that time, roads were developed and forests cleared to establish human settlements and oil palm plantations. One of the only surviving fragments is the 1.7 km^2 Rio Palenque Science Reserve. This tiny conservation area has 1025 recorded plant species, of which 25% are not known to occur anywhere else (Gentry 1986). Over 100 undescribed plant species have been recorded at this site. Many of these species are known only from a single individual plant and are doomed to certain extinction.

Other threatened habitats. The plight of the rain forest is perhaps the most widely publicized case of habitat destruction, but other habitats are also in grave danger.

- *Tropical dry forests.* The land occupied by tropical dry forests is more suitable for agriculture and cattle ranching than is the land occupied by tropical rain forests. Consequently, human popula- tion density is five times greater in dry forest areas of Central America than in adjacent rain forest areas (Murphy and Lugo 1986). Today, the Pacific Coast of Central America has less than 2% of its original extent of deciduous dry forest remaining (Janzen 1988a).

- *Wetlands and aquatic habitats.* Wetlands are critical habitats for fish, aquatic invertebrates, and birds. They are also a resource for flood control, drinking water, and power production (Mitchell 1992; Dugan 1993). Wetlands are often filled in or drained for development or altered by restricting meandering watercourses into artificial channels, the creation of dams, and chemical pollution. All of these factors are affecting the Florida Everglades, one of the prime wildlife refuges in the United

States, which is now on the verge of ecological collapse, but is also the target of massive recovery projects.

During the last 200 years, over half of the wetlands in the United States have been destroyed, resulting in 60%–70% of the fresh-water mussel species in the United States becoming either extinct or endangered (Stein and Flack 1997). Destruction of wet-lands has been equally severe in other areas of the industrialized world, such as Europe and Japan. Throughout the world, most salmon populations are in decline because of dams, which pre-vent migration up and down river channels. In the last few decades, one of the major threats to wetlands in developing countries has been massive development projects involving drainage, irrigation, and dams; such projects are organized by governments and often financed by international aid agencies. Although many wetland species are widespread, some aquatic systems are known for their high levels of endemism. For exam-ple, Lake Victoria in East Africa has one of the richest endemic fish faunas in the world, but 250 of its species are in danger of extinction due to water pollution and introductions of exotic fishes that prey on the endemic fishes (Kaufman 1992).

- *Mangroves.* One of the most important wetland communities in tropical areas is the mangrove forest. Mangrove species are among the few woody plants that can tolerate salt water. Man-grove forests occupy coastal areas with saline or brackish water, typically where there are muddy bottoms. Such habitats are sim-ilar to those occupied by salt marshes in the temperate zone. Mangrove forests are extremely important as breeding grounds and feeding areas for shrimp and fish. In Australia, two-thirds of the species caught by commercial fishermen are dependent to some degree on the mangrove ecosystem. Despite their great economic value, mangrove forests are often cleared for rice culti-vation and commercial shrimp and prawn hatcheries, particu-larly in Southeast Asia. Mangroves are also a source of wood for poles, charcoal, and industrial production, often leading to over-collection and habitat degradation. The loss of mangroves is extensive in South and Southeast Asia; it is particularly high for India (85%), Thailand (87%), Pakistan (78%), and Bangladesh (73%) (WRI 1994).

- *Grasslands.* Temperate grassland is another habitat type that has been almost completely destroyed by human activity. It is rela-tively easy to convert large areas of grassland into farmland and cattle ranches. Illinois and Indiana originally contained 15 million ha of tall-grass prairie, but now only 1400 ha of this habitat—one ten-thousandth of the original area—remains undisturbed; the rest has been converted into farmland (Chadwick 1993). This remaining area of prairie is divided up into many small fragments, which are widely scattered across the landscape.

- *Coral reefs.* Tropical coral reefs contain an estimated one-third of the ocean's fish species, although they occupy only 0.2% of its surface area. Already, 10% of all coral reefs have been destroyed, and as many as 50% could be destroyed in the next few decades (Birkeland 1997). The most severe destruction is taking place in the Philippines, where a staggering 90% of reefs are dead or dying. The main culprits are pollution, which either kills the coral directly or allows excess growth of algae; sedimentation following the removal of forests; and overharvesting of fish, clams, and other animals. In some areas, methods such as blasting with dynamite and releasing cyanide are used by fishermen to stun and then collect the remaining living creatures. Extensive loss of coral reefs is expected within the next 40 years in tropical East Asia, around Madagascar and East Africa, and throughout the Caribbean (Figure 2.12).

Desertification. Many biological communities in seasonally dry climates have been degraded into artificial deserts by human activities, a process known as **desertification** (Allan and Warren 1993). Such communities include tropical grasslands, scrub, and deciduous forest, as well as temperate shrublands and grasslands such as those found in the Mediterranean region, southwestern Australia, South

2.12 Extensive areas of coral will be damaged or destroyed by human activity over the next 40 years unless conservation measures can be implemented. (After Bryant et al. 1998.)

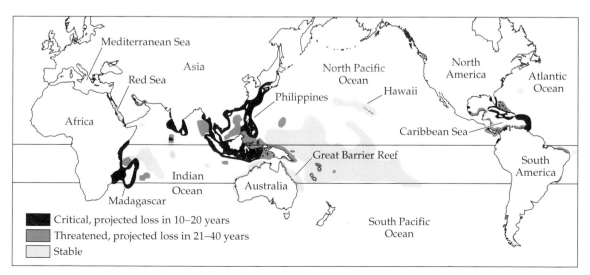

2.13 A grassland in Utah, overgrazed by cattle, takes on the appearance of a desert, leading to the elimination of native species. (Photograph © Rod Planck, Photo Researchers, Inc.)

Africa, Chile, and southern California. While these areas initially may be suitable for agriculture, repeated cultivation leads to soil erosion and a loss of the water-holding capacity of the soil. Land may also be chronically overgrazed by domestic livestock such as cattle, sheep, and goats (Figure 2.13), and woody plants may be cut down for fuel (Fleischner 1994; Milton et al. 1994). The result is a progressive and largely irreversible degradation of the biological community and a loss of soil cover, to the point at which the area takes on the appearance of a desert. Worldwide, it has been estimated that 9 million km² of arid lands have been converted into deserts by this process (Dregné 1983). Desertification is particularly severe in the Sahel region of Africa, an area on the southern edge of the Sahara ranging from Mauritania to Chad, where most of the native large mammal species are threatened with extinction. The human dimensions of the problem are illustrated by the fact that the Sahel region is estimated to have 2.5 times more people than the land can sustainably support without further intensification of agriculture.

Habitat fragmentation

In addition to being destroyed outright, habitats that formerly occupied large areas are often divided into small pieces by roads, fields, towns, and a wide range of other human constructs. **Habitat fragmentation** is the process whereby a large, continuous area of habitat is both reduced in area and divided into two or more fragments (Shafer 1990; Reed et al. 1996). When habitat is destroyed, a patchwork of habitat fragments may be left behind. These fragments are often isolated from one another by a highly modified or degraded

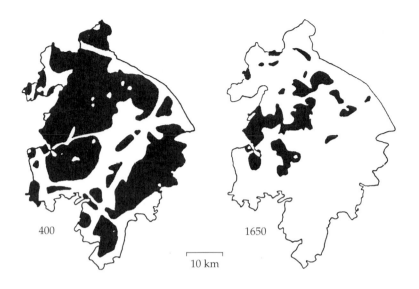

400   1650   1960

10 km

2.14 The forested areas of Warwickshire, England (shown in black), have been fragmented and reduced in area—by paths, roads, agriculture, and human settlements—over the centuries from 400 (when Romans established towns in the forested landscape and built roads between them) to 1960 A.D. (when only a few tiny forest fragments remain). (From Wilcove et al. 1986.)

landscape (Figure 2.14). As mentioned earlier, this situation can be described by the island model of biogeography, with the fragments functioning as habitat islands in an inhospitable human-dominated "sea." Fragmentation occurs during almost any severe reduction in habitat area, but it can also occur even when habitat area is reduced to only a minor degree, as when the original habitat is divided by roads, railroads, canals, power lines, fences, oil pipelines, fire lanes, or other barriers to the free movement of species.

Habitat fragments differ from the original habitat in two important ways: (1) fragments have a greater length of **edge habitat** (habitat adjacent to human activities) per area of total habitat, and (2) the center of each habitat fragment is closer to an edge. A simple example will illustrate these characteristics and the problems that can occur because of them.

Consider a square conservation reserve 1000 m (1 km) on each side and completely surrounded by human-dominated lands, such as farms (Figure 2.15). The total area of the reserve is 1 km^2 (100 ha); the perimeter (or edge) of the reserve totals 4000 m, and a point in the middle of the reserve is 500 m from the nearest point on the perimeter. If domestic cats forage 100 m into the forest from the perimeter of the reserve and prevent forest birds from successfully raising their young, then only the 64 ha in the reserve's interior is available to the birds for breeding. Edge habitat, unsuitable for breeding, occupies 36 ha.

Now imagine the reserve being divided into four equal quarters by a north–south road 10 m wide and by an east–west railroad track

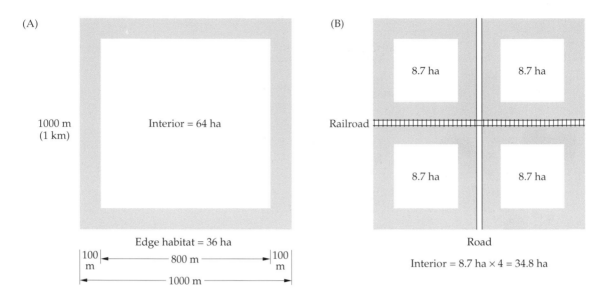

(A)

1000 m
(1 km)

Interior = 64 ha

Edge habitat = 36 ha

100 m ←——— 800 m ———→ 100 m

←————— 1000 m —————→

(B)

Railroad

8.7 ha   8.7 ha

8.7 ha   8.7 ha

Road

Interior = 8.7 ha × 4 = 34.8 ha

2.15 This hypothetical example shows how habitat area is reduced by fragmentation and edge effects. (A) This 1 km² protected area is not fragmented. Assuming that edge effects (gray) penetrate 100 m into the reserve, approximately 64 ha are available as usable interior habitat for nesting birds. (B) The bisection of the reserve by a road and a railway, although they take up little in actual area, extends the edge effects so that almost half the nesting habitat is destroyed.

that is also 10 m wide. The rights-of-way remove a total of 2 × 1000 m × 10 m of area (2 ha) from the reserve. Because only 2% of the reserve is being removed by the road and railroad, government planners argue that the effects on the reserve are negligible. However, the reserve has now been divided into four fragments, each of which is 495 m × 495 m in area, and the distance from the center of each fragment to the nearest point on the perimeter has been reduced to 247 m, which is less than half of the former distance. Because cats can now forage into the forest from the road and railroad as well as the perimeter, birds can successfully raise young only in the interior areas of each of the four fragments. Each of these interior areas is 8.7 ha, for a total of 34.8 ha. Even though the road and railroad removed only 2% of the reserve area, they reduced the habitat available to the birds by about half.

Habitat fragmentation also threatens the persistence of species in more subtle ways. First, fragmentation may limit a species' potential for dispersal and colonization. Many bird, mammal, and insect species of the forest interior will not cross even very short stretches of open area because of the danger of predation. As a result, many species do not recolonize fragments after the original population has disappeared (Laurance and Bierregaard 1997). Furthermore, when animal dispersal is reduced by habitat fragmentation, plants with fleshy fruits or sticky seeds that depend on the animals to disperse their seeds will be affected as well. Thus, isolated habitat fragments will not be colonized by many native species that could potentially live in them. As species become extinct within individual fragments through natural successional and population processes, new species will fail to arrive due to these dispersal barriers, and the number of species in the habitat fragment will decline over time.

A second harmful aspect of habitat fragmentation is that it may reduce the foraging ability of native animals. Many animal species, either as individuals or as social groups, need to be able to move freely across the landscape to feed on widely scattered or seasonally available food resources and to find water. A given resource may be needed for only a few weeks of the year, or even only once in a few years, but when a habitat is fragmented, species confined to a single habitat fragment may be unable to migrate over their normal home range in search of that scarce resource. For example, fences may prevent the natural migration of large grazing animals such as wildebeest or bison, forcing them to overgraze an unsuitable habitat and eventually leading to starvation of the animals and degradation of the habitat.

Habitat fragmentation may also precipitate population decline and extinction by dividing an existing widespread population into two or more subpopulations, each in a restricted area (Rochelle et al. 1999). These smaller populations are more vulnerable to inbreeding depression, genetic drift, and other problems associated with small population size (see Chapter 3). While a large area of habitat may have supported a single large population, it is possible that none of its fragments can support a subpopulation large enough to persist for a long period.

Edge effects. Habitat fragmentation dramatically increases the amount of edge relative to the amount of interior habitat, as demonstrated above (see Figure 2.15). The microenvironment at a fragment edge is different from that of the forest interior. Some of the more important **edge effects** are greater fluctuations in levels of light, temperature, humidity, and wind (Figure 2.16) (Schelhas and Greenberg 1996; Laurance and Bierregaard 1997). These edge effects are often evident up to 250 meters into the forest. Because plant and animal species are often precisely adapted to certain temperature, humidity, and light levels, these changes will eliminate many species from forest fragments. Shade-tolerant wildflower species of the temperate forest, late-successional tree species of the tropical forest, and humidity-sensitive animals such as amphibians are often rapidly eliminated by habitat fragmentation, leading to a shift in the species composition of the community.

When a forest is fragmented, the increased wind, lower humidity, and higher temperatures at the forest edge make fires more

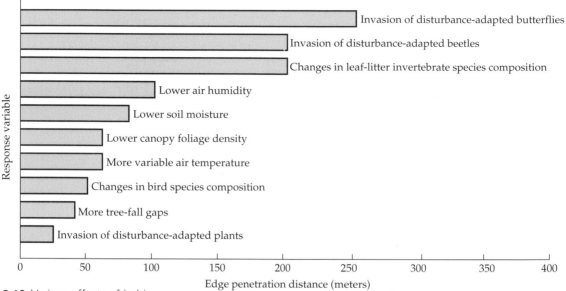

2.16 Various effects of habitat fragmentation, as measured from the edge into the interior of an Amazon rain forest fragment. The bars indicate how far into the forest the edge effect occurs. For example, disturbance-adapted butterflies migrate 250 m into the forest from an edge, and the relative humidity of the air is lowered within 100 m of the forest edge. (After Laurance and Bierregaard 1997.)

likely. Fires may spread into habitat fragments from nearby agricultural fields that are burnt regularly, as in sugarcane harvesting, or from the activities of farmers practicing shifting cultivation. In Borneo and the Brazilian Amazon, millions of hectares of tropical moist forest burned during an unusual dry period in 1997 and 1998; a combination of forest fragmentation due to farming and selective logging, accumulation of brush after selective logging, and human-caused fires contributed to these environmental disasters (Goldammer 1999; Cochrane et al. 1999).

Habitat fragmentation increases the vulnerability of the fragments to invasion by exotic species and native pest species. The forest edge is a disturbed environment in which pest species can easily become established, increase in numbers, and then disperse into the interior of the fragment (Paton 1994). Omnivorous animals such as raccoons, skunks, and blue jays may increase in numbers along forest edges, where they can obtain foods from both undisturbed and disturbed habitats. These aggressive feeders eat the eggs and nestlings of forest birds, often preventing successful reproduction for many bird species hundreds of meters from the nearest forest edge. Nest-parasitizing cowbirds (*Molothrus ater*), which live in fields and edge habitats, use habitat edges as an invasion point into forest interiors, where their nestlings destroy the eggs and nestlings of forest songbirds. The combination of habitat fragmentation, increased nest predation, and destruction of tropical wintering habitats is probably responsible for the dramatic decline of certain migratory songbird species of North America, such as the cerulean warbler (*Dendroica cerulea*), particularly in the eastern half of the United States. In addition to these local effects, individual bird species in North America and Europe are both increasing and decreasing on regional scales in response to changing land use patterns, such as agricultural practices and forest management activities (James et al 1996; Fuller et al. 1995).

Habitat fragmentation also brings wild populations into contact with domestic plants and animals. Diseases of domestic species can then spread more readily to wild species, which often have low immunity to them. There is also a potential for diseases to spread from wild species to domestic plants, animals, and even people, once the level of contact increases.

Habitat degradation and pollution

Even when a habitat is unaffected by overt destruction or fragmentation, the communities and species in that habitat can be profoundly affected by human activities. Biological communities can be damaged and species driven to extinction by external factors that do not change the dominant plant structure of the community, so that the damage is not immediately apparent. For example, in a temperate deciduous forest, physical degradation of a habitat might be caused by frequent uncontrolled ground fires; these fires might not kill the mature trees, but the rich perennial wildflower community and insect fauna on the forest floor would be gradually eliminated. Out of sight from the public, fishing trawlers drag across an estimated 15 million km^2 of ocean floor each year, an area 150 times greater than the area of forest cleared in the same time period. The trawling destroys delicate creatures such as anemones and sponges and reduces species diversity, biomass, and community structure (Figure 2.17) (Watling and Norse 1998).

The most subtle and universal form of environmental degradation is environmental pollution, the most common causes of which are pesticides, chemicals and sewage released by industries and human settlements, emissions from factories and automobiles, and sediment deposits from eroded hillsides. These types of pollution are often not visually apparent even though they occur all around us, every day, in nearly every part of the world. The general effects of pollution on water quality, air quality, and even the global climate are cause for great concern, not only as threats to biological diversity but also because of their effects on human health. Although environmental pollution is sometimes highly visible and dramatic, as in the case of the massive oil spills and 500 oil well fires that resulted from the Persian Gulf War, it is the subtle, unseen

(A)

(B)

2.17 Undisturbed gravel beds in Georges Bank, off south-western Nova Scotia, are occupied by dense colonies of tube worms, hydrozoans, bryozoans, and other marine animals (A). Five hundred meters away, gravel beds that have been chronically dragged for scallops support comparatively little animal life, just a few bryozoans, anemones, and empty scallop shells (B). The field of view in both photos is 35 cm in length. (Photographs by the U.S. Geological Survey, courtesy of Page Valentine.)

forms of pollution that are probably the most threatening—primarily because they are so insidious.

Pesticide pollution. The dangers of pesticides were brought to the world's attention in 1962 by Rachel Carson's influential book *Silent Spring*. Carson described the process known as **biomagnification**, through which DDT (dichloro-diphenyltrichloroethene) and other organochlorine pesticides become more concentrated as they ascend the food chain. These pesticides, used on crop plants to kill insects and sprayed on bodies of water to kill mosquito larvae, were harming wildlife populations, particularly birds that ate large amounts of insects, fish, or other animals exposed to DDT and its by-products. Birds with high levels of pesticides concentrated in their tissues, particularly raptors such as hawks and eagles, were weakened and tended to lay eggs with abnormally thin shells, which cracked during incubation. Even when the egg shells were normal, embryo development was often defective. As a result of their failure to raise young, populations of these birds showed dramatic declines throughout the world. In lakes and estuaries, DDT and other pesticides became concentrated in predatory fishes and marine mammals such as dolphins. In agricultural areas, beneficial and endangered insect species were killed along with harmful species. At the same time, mosquitoes and other targeted insects evolved resistance to the chemicals, so that ever-larger doses of DDT were required to suppress the insect populations. Recognition of this situation led to a ban on the use of organochloride pesticides in many industrialized countries. The ban eventually allowed the partial recovery of many bird species, most notably peregrine falcons (*Falco peregrinus*), ospreys (*Pandion haliaetus*), and bald eagles

(*Haliaetus leucocephalus*) (Enderson et al. 1995). Nevertheless, the continuing use of these pesticides in other countries is cause for concern, not only for the sake of endangered animal species, but also because of their potential long-term effects on people, particularly workers who handle the chemicals in the field and people who eat agricultural products treated with them. In addition, these chemicals persist in the environments of countries that banned them decades ago; there they continue to have a detrimental effect on the reproductive systems of aquatic invertebrates (McLachlan and Arnold 1996).

Water pollution. Water pollution has negative consequences for human populations: it destroys food sources such as fish and shell-fish and contaminates drinking water. On a broader scale, water pollution often severely damages aquatic communities (Figure 2.18). Pollution is a threat to 90% of the endangered fishes and freshwater mussels in the United States (Wilcove et al. 1998). Rivers,

2.18 Aquatic habitats such as rivers, lakes, estuaries, and the open ocean are often used as dumping grounds for sewage, rubbish, and industrial wastes, to the detriment of biological communities. (After Eales 1992.)

lakes, and oceans are often used as open sewers for industrial wastes and residential sewage. Pesticides, herbicides, petroleum wastes and spills, heavy metals (such as mercury, lead, and zinc), detergents, and industrial wastes can injure and kill organisms living in aquatic environments. An increasing source of pollution in coastal areas is the discharge of nutrients and chemicals from shrimp and salmon farms (Naylor et al. 1998). In contrast to wastes dumped in the terrestrial environment, which have primarily local effects, toxic wastes in aquatic environments can be carried by currents and diffused over a wide area. Toxic chemicals, even in very small amounts, may be concentrated to lethal levels by aquatic organisms that filter large volumes of water while feeding. Bird and mammal species that prey on the filter feeders are exposed in turn to these concentrated levels of toxic chemicals.

Even essential minerals that are beneficial to plant and animal life can become harmful pollutants at high levels. At present, human activities release as much nitrogen into biological communities each year as are taken up by natural biological processes. Human sewage, agricultural and lawn fertilizers, detergents, and industrial processes often release large amounts of nitrogen and phosphorus compounds into aquatic systems, starting a process known as **cultural eutrophication**. Although small amounts of these nutrients can stimulate plant and animal growth, high concentrations often result in thick "blooms" of algae at the water's surface. These algal blooms may be so dense that they outcompete other plankton species and shade out bottom-dwelling plant species. As the algal mat becomes thicker, its lower layers sink to the bottom and die. The bacteria and fungi that decompose the dying algae multiply in response to this added sustenance and consequently absorb all of the oxygen in the water. Without oxygen, much of the remaining animal life dies off, sometimes visibly in the form of masses of dead fish floating on the water's surface. The result is a greatly impoverished and simplified community, consisting only of species that can tolerate polluted water and low oxygen levels. This process of eutrophication can affect large marine systems as well, particularly coastal areas and bodies of water in a confined area, such as the Gulf of Mexico, the North Sea and the Baltic Sea in Europe, and the enclosed seas of Japan (Malakoff 1998).

Eroding sediments from logged or farmed hillsides can also harm aquatic ecosystems. The sediments cover submerged plant leaves and other green surfaces with a muddy film that reduces light availability and diminishes the rate of photosynthesis. Increasing water turbidity may prevent animal species from seeing, feeding, and living in the water, and it may reduce the depth at which light penetrates and photosynthesis can occur. Increased sediment loads are particularly harmful to many coral species, which require crystal-clear water to survive.

Air pollution. In the past, people assumed that the atmosphere was so vast that materials released into the air would be widely dispersed and their effects would be minimal. But today several types of air pollution are so widespread that they damage whole ecosystems.

- *Acid rain.* Industries such as smelting operations and coal- and oil-fired power plants release huge quantities of nitrogen and sulfur oxides into the air, where they combine with moisture in the atmosphere to produce nitric acid and sulfuric acid. In the United States alone, around 40 million metric tons of these compounds are released into the atmosphere each year (WRI 1998). The acids are incorporated into cloud systems and dramatically lower the pH (the standard measure of acidity) of rainwater. Acid rain in turn lowers the pH of soil moisture and water bodies such as ponds and lakes. By itself, the acidity is damaging to many plant and animal species. As the acidity of water bodies is increased by acid rain, many fish either fail to spawn or die outright (Figure 2.19). In addition to introduced predators, changing weather patterns, increased ultraviolet light levels, and the spread of exotic diseases, increased acidity and water pollution are two likely factors behind the recent dramatic decline in amphibian populations throughout the world. Most amphibian species depend on bodies of water for at least part of their life cycle, and a decline in water quality causes a corresponding increase in the mortality of eggs and larvae both directly and via increased susceptibility to disease (Blaustein and Wake 1995; Halliday 1998; Alford and Richards 1999). Many ponds and lakes in industrialized areas of the world have lost large por-

2.19 The pH scale, indicating ranges at which acidity becomes lethal to fish. Studies indicate that fish are indeed disappearing from heavily acidified lakes. (After Cox 1993; based on data from the U.S. Fish and Wildlife Service.)

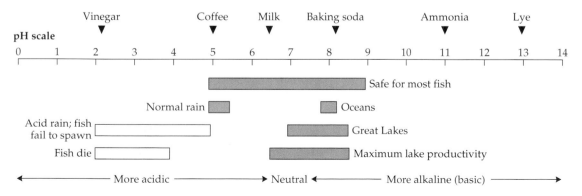

tions of their animal communities as a result of acid rain. Many of these damaged water bodies are in supposedly pristine areas hundreds of kilometers from major sources of urban and industrial pollution; for instance, acidification is already evident in 39% of lakes in Sweden and 34% of lakes in Norway (Moiseenko 1994). While acidity of rain is decreasing in many areas due to better pollution control, it remains far above normal levels (Kerr 1998).

- *Ozone production and nitrogen deposition.* Automobiles, power plants, and other industrial activities release hydrocarbons and nitrogen oxides as waste products. In the presence of sunlight, these chemicals react with the atmosphere to produce ozone and other secondary chemicals, collectively called **photochemical smog**. Although ozone in the upper atmosphere is important in filtering out harmful ultraviolet radiation, high concentrations of ozone at ground level damage plant tissues and make them brittle, harming biological communities and reducing agricultural productivity. Biological communities throughout the world can also be damaged and altered when these airborne nitrogen compounds are deposited by rain and dust, leading to potentially toxic levels of this nutrient. Ozone and smog are detrimental to both people and animals when inhaled, so controlling air pollution benefits humans and biological diversity.

- *Toxic metals.* Leaded gasoline, mining and smelting operations, and other industrial activities release large quantities of lead, zinc, and other toxic metals into the atmosphere. These compounds are directly poisonous to plant and animal life. The effects of these toxic metals are particularly evident in areas surrounding large smelting operations, where life can be destroyed for miles around.

The effects of air pollution on forest communities have been intensively studied because forests have such great economic value in terms of wood production, the protection of water supplies ("watershed management"), and recreation. It is widely accepted that acid rain damages and weakens many tree species and makes them more susceptible to attacks by insects, fungi, and disease (Figure 2.20). Widespread deaths of forest trees over large areas of Europe and eastern North America have been linked to acid rain and other components of air pollution such as nitrogen deposition and ozone. When the trees die, other species in the forest also become extinct on a local scale. Even when communities are not destroyed by air pollution, species composition may be altered as more susceptible species are eliminated. Lichens—symbiotic organisms composed of fungi and algae that can survive in some of the

2.20 Forests throughout the world are experiencing diebacks, thought to be caused in part by the effects of acid rain, nitrogen deposition, and ozone, leaving trees more vulnerable to damage by insect attack, disease, and drought. These dead trees were photographed on Mt. Mitchell, North Carolina, in 1988. (Photograph by Jim MacKenzie, WRI.)

harshest natural environments—are particularly susceptible to air pollution.

Levels of air pollution are declining in certain areas of North America and Europe, but continue to rise in many other areas of the world. Increases in air pollution will be particularly severe in the many Asian countries with dense (and growing) human populations and increasing industrialization. The heavy reliance of China on high-sulfur coal and the rapid increase in automobile ownership in Southeast Asia are examples of potential threats to biological diversity in those regions, with the production of sulfur dioxide projected to double between 2000 and 2020 (WRI 1998). Hope for controlling air pollution in the future depends on motor vehicles with dramatically lower emissions of pollutants, increased development of mass transit systems, more efficient scrubbing processes for industrial smokestacks, and the reduction of overall energy use through conservation and efficiency measures. Many of these measures are already being actively implemented in European countries and Japan.

Global climate change. Carbon dioxide, methane, and other trace gases in the atmosphere are transparent to sunshine, allowing light energy to pass through the atmosphere and warm the surface of the Earth. However, these gases and water vapor (seen in the form of clouds) trap the energy radiating from the Earth's surface as heat, slowing the rate at which heat leaves the Earth and radiates back into space. These gases are called **greenhouse gases** because they function much like greenhouse glass, which is transparent to sunlight but traps energy inside the greenhouse once it is transformed into heat (Figure 2.21). The denser the concentration of gases, the more heat is trapped near the Earth, and the higher the planet's surface temperature. This is called the **greenhouse effect**.

The greenhouse effect has been important in allowing life to flourish on Earth—without it, the temperature at the Earth's surface would fall dramatically and life as we know it would not be possible. The problem that exists today, however, is that concentrations of greenhouse gases are increasing so much as a result of human activity that scientists believe they are already affecting the Earth's climate. The term **global warming** is used to describe this enhanced greenhouse effect resulting from human activities. During the past 100 years, global atmospheric levels of carbon dioxide (CO_2), methane, and other trace gases have been steadily increasing, pri-

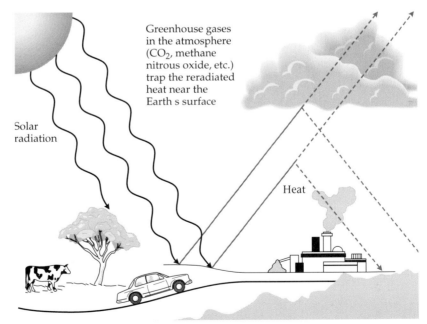

Greenhouse gases in the atmosphere (CO_2, methane nitrous oxide, etc.) trap the reradiated heat near the Earth s surface

Solar radiation

Heat

2.21 In the greenhouse effect, gases and water vapor form a blanket around the Earth that acts like the glass roof of a greenhouse, trapping heat near the Earth's surface. (From Gates 1993.)

Light waves are transformed into infrared radiation (heat) and reradiated

marily as a result of the burning of fossil fuels such as coal, oil, and natural gas (Gates 1993; IPCC 1996). The cutting down and burning of forests to create farmland and the burning of fuelwood for heating and cooking also contribute to rising concentrations of CO_2. The CO_2 concentration in the atmosphere has increased from 290 parts per million (ppm) to 360 ppm over the last 100 years, and it is projected to double somewhere in the latter half of the twenty-first century. Even if immediate, massive efforts are made to reduce CO_2 production, there will be little immediate reduction in present atmospheric CO_2 levels because each CO_2 molecule resides in the atmosphere for an average of 100 years before being removed by plants and natural geochemical processes. Because of this time lag, levels of CO_2 in the atmosphere will continue to rise.

Many scientists believe that increased levels of greenhouse gases have affected the world's climate already, and that these effects will increase in the future (Table 2.7). An extensive review of the evi-

TABLE 2.7   Some evidence for global warming

1. Increased incidence of heat waves

Example: A July 1999 heat wave in the United States kills 250 people; Chicago registers a record temperature of 48°C (119°F).

2. Increased incidence of droughts and fires

Example: Severe summer droughts in 1998 are followed by massive wildfires in Indonesia, Central America, southern Europe, and the southern United States.

3. Melting of glaciers and polar ice

Example: In the Caucasus Mountains between the Black Sea and the Caspian Sea, half of all glacial ice has melted during the last 100 years.

Example: A 2992 km^2 section of previously stable Antarctic ice shelf collapses in 1999.

4. Rising sea levels

Example: Since 1938, one-third of the coastal marshes in a wildlife refuge in Chesapeake Bay has been submerged by rising seawater.

5. Spread of disease to higher elevations

Example: In 1997, rising temperatures allowed malaria-carrying mosquitoes to extend their range into the Kenya highlands, killing hundreds of people.

6. Earlier spring arrival

Example: One-third of English birds are now laying eggs earlier in the year then they did 30 years ago, and oak trees are now leafing out earlier than they did 40 years ago.

7. Shifts in species ranges

Example: Two-thirds of European butterfly species studied are now found farther northward by 35 to 250 km than recorded several decades ago.

8. Population declines

Example: Adelie penguin populations have declined by one-third over the past 25 years as their sea ice habitat melts away.

Source: After Union of Concerned Scientists 1999.

dence supports the conclusion that world climate has warmed by between 0.3° and 0.6° Celsius (°C) during the last century (IPCC 1996; Schneider 1998). Recent evidence indicates that ocean water temperatures are also changing: Over the last 50 years, the Atlantic, Pacific, and Indian Oceans have increased by an average of 0.06° C (Levitus et al. 2000). There is a growing consensus among meteorologists that the world climate will probably warm by an additional 1° to 3.5°C over the next century as a result of increased atmospheric levels of carbon dioxide and other gases. Associated with this warming will be altered patterns of precipitation and an increased incidence of extreme weather events, such as hurricanes, heat waves, heavy rainstorms leading to flooding, and droughts. It seems likely that many species will be unable to adjust quickly enough to survive this human-engendered global change, which will occur far more rapidly than previous natural climate shifts.

While the details of global climate change are still being debated by scientists, there is no doubt that the effects of such a rapid rise in temperature on biological communities will be profound. For example, climatic regions in the northern and southern temperate zones will be shifted toward the poles. Species adapted to the eastern deciduous forests of North America will have to migrate up to 500–1000 km northward over the twenty-first century to keep up with this climate shift (Davis and Zabinski 1992). Although widely distributed, easily dispersed species, such as many butterflies and migratory birds, may be able to adjust to these changes, many species with limited distributions or poor dispersal ability will undoubtedly become extinct. These extinctions will be accelerated by habitat fragmentation, which creates barriers to dispersal. Species currently restricted to isolated mountain peaks and fish species found in a single lake or watershed will be particularly vulnerable to extinction. Our current system of national parks and nature reserves will no longer provide protection for the species and ecosystems that it was designed to safeguard.

Warming temperatures are apparently already melting mountain glaciers and shrinking the polar ice caps. As a result of this release of water over the next 50 to 100 years, sea levels could rise by 0.2–1.5 meters. This rise in sea level will flood low-lying coastal wetland communities and many major urban areas. Particularly where human settlements, roads, and flood control barriers have been built adjacent to wetlands, the migration of wetland species will be blocked. It is possible that rising sea levels could destroy or alter 25%–80% of the coastal wetlands of the United States. In low-lying countries such as Bangladesh, much of the land area could be under water within 100 years. There is already evidence that this process has begun; sea levels have already risen by 10–25 cm over the last 100 years, probably due to rising global temperatures.

Rising sea levels will be potentially detrimental to many coral species, which grow at a precise depth in the water that has the right combination of light and current. Certain coral reefs may be unable to grow quickly enough to keep pace with the rise in sea level and will gradually be "drowned." Damage could be compounded if ocean temperatures also rise (Wilkinson et al. 1999). Abnormally high water temperatures in the Indian Ocean and the Pacific Ocean in 1998 led to the death of the symbiotic algae that live inside corals; these "bleached" corals then suffered a massive dieback, with an estimated 70% coral mortality in the Indian Ocean reefs.

Global climate change and increasing concentrations of atmospheric CO_2 have the potential to radically restructure biological communities, favoring those species that are able to adapt to the new conditions (Bazzaz and Fajer 1992). There is mounting evidence that this process of change has already begun (see Table 2.7), with pole-ward movements in the distribution of bird and butterfly species, and earlier reproduction in the spring (Thomas and Lennon 1999; Parmesan et al. 1999). Because the implications of global climate change are so far-reaching, biological communities, ecosystem functions, and climate need to be carefully monitored in the future. To help species and biological communities survive, conservation biologists will have to assist in establishing new parks with large elevational gradients and north–south migration routes. Another strategy will be to transplant isolated populations of rare and endangered species to new localities at higher elevations and farther from the poles where they can survive and thrive.

Concerns over global climate change, however, should not divert attention from the massive habitat destruction that is still the principal cause of species extinction: Saving intact communities from destruction and restoring degraded communities are still the most important and immediate priorities for conservation. It is also important to emphasize that global climate change will have enormous impacts on the world economy. Agriculture areas will have to be relocated, huge dikes will have to be built to protect coastal cities, and increased storms, heat waves, and droughts will affect people everywhere in the world. And as is often the case, poor people in developing countries will be affected most severely by these changes when their crops fail, their houses wash away, their wells run dry, and new diseases kill their children. Both the world's biological communities and its human populations will pay in the end for our inability to act decisively on the issue of global climate change.

Overexploitation

People have always hunted and harvested the food and other resources they needed to survive. As long as human populations were small and their methods of collection unsophisticated, people could sustainably harvest and hunt the plants and animals of their environment without driving species to extinction. However, as human populations have increased, their use of the environment has escalated. Methods of harvesting have also become dramatically more efficient, leading to an almost complete depletion of large mammals from many biological communities, resulting in strangely "empty" habitats (Redford 1992). Guns are used instead of blowpipes, spears, or arrows for hunting in the tropical rain forests and savannahs. Powerful motorized fishing boats and efficient "factory ships" harvest fish from all the world's oceans. Small-scale local fishermen have outboard motors on their canoes and boats, allowing them to harvest more rapidly and over a wider area than previously possible. Even in preindustrial societies, intense exploitation has led to the decline and extinction of local species. For example, ceremonial cloaks worn by the Hawaiian kings were made from the feathers of the mamo bird (*Drepanis* sp.); a single cloak used the feathers of 70,000 birds of this now-extinct species. Predator species may also decline when their prey are overharvested by people. Overexploitation by humans is estimated to threaten about a quarter of the endangered vertebrates in the United States, and around half of its endangered mammals (Wilcove et al. 1998; Wilcove 1999).

In traditional societies, restrictions often existed to prevent the overexploitation of natural resources: for example, the rights to specific harvesting territories were rigidly controlled; hunting in certain areas was banned; there were prohibitions against killing females, juveniles, and undersized individuals; certain seasons of the year and times of the day were closed for harvesting; or certain efficient types of harvesting were not allowed. These kinds of restrictions, which allowed traditional societies to harvest communal resources on a long-term, sustainable basis, are very similar to the rigid fishing restrictions developed and proposed for many fisheries in industrialized nations (Freese 1997).

In much of the world today, however, resources are exploited as rapidly as possible. If a market exists for a product, local people will search their environment to find and sell it. Whether people are poor and hungry or rich and acquisitive, they will use whatever methods are available to secure the product. Sometimes traditional societies even decide to sell their rights to a resource, such as a forest or mining area, in order to use the money to buy desired or urgently needed goods. In rural areas, the traditional controls that regulated the extraction of natural products have generally weakened, and in many areas into which there has been substantial human migration, or where civil unrest and war have occurred, such controls may no longer exist at all. In countries beset with civil conflict, such as Somalia, the former Yugoslavia, the Democratic

Republic of the Congo, and Rwanda, there has been a proliferation of firearms among rural people and a breakdown of food distribution networks. In such situations, the resources of the natural environment will be taken by whoever can exploit them. On local and regional scales, hunters in developing countries move into recently logged areas, national parks, and other areas near roads and remove every large animal to sell as wild meat, creating an "empty forest": land with a mostly intact plant community but lacking its animal community (Robinson et al. 1999).

Overexploitation of resources often occurs rapidly when a commercial market develops for a previously unexploited or locally used species. One of the most pervasive examples is the international trade in furs, which has reduced species such as the chinchilla (*Chinchilla* spp.), vicuña (*Vicugna vicugna*), giant otter (*Pteronura brasiliensis*), and numerous cat species to very low numbers. The legal and illegal trade in wildlife is responsible for the decline of many species (Poten 1991; Hemley 1994). Overharvesting of butterflies by insect collectors, of orchids, cacti, and other plants by horticulturists, of marine mollusks by shell collectors, and of tropical fishes for aquarium hobbyists are further examples in which whole biological communities have been targeted to supply an enormous international demand (Table 2.8).

TABLE 2.8 Major groups targeted by the worldwide trade in wildlife

Group	Number traded each year[a]	Comments
Primates	25–30 thousand	Mostly used for biomedical research; also for pets, zoos, cir cuses, and private collections
Birds	2–5 million	Zoos and pets. Mostly perching birds, but also legal and illegal trade in parrots.
Reptiles	2–3 million	Zoos and pets. Also 10–15 million raw skins. Reptiles are used in some 50 million manufactured products (mainly coming from the wild but increasingly from farms).
Ornamental fish	500–600 million	Most saltwater tropical fish come from the wild and may be caught using illegal methods that damage other wildlife and the surrounding coral reef.
Reef corals	1000–2000 tons	Reefs are being destructively mined to provide aquarium decor and coral jewelry.
Orchids	9–10 million	Approximately 10% of the international trade comes from the wild, sometimes deliberately mislabeled to avoid regulations.
Cacti	7–8 million	Approximately 15% of the traded cacti come from the wild, with smuggling a major problem.

Source: Data from Hemley 1994 and Fitzgerald 1989.

[a]With the exception of reef corals, refers to number of individuals.

The pattern in many cases of overexploitation is distressingly familiar. A resource is identified, a commercial market is developed for that resource, and the local human populace mobilizes to extract and sell the resource. The resource is extracted so thoroughly that it becomes rare or even extinct, and the market identifies another species or another region to exploit in place of the first. Commercial fishing fits this pattern well, with the industry working one species after another to the point of diminishing return. Commercial forestry companies often behave similarly, with loggers extracting less desirable tree species or trees of smaller size in successive cutting cycles until there is little timber left in the forest, at which time the companies move to another unexploited location. Hunters move successively farther from their villages and logging camps to find animals to catch for their own consumption and for sale.

An extensive body of literature has developed in the fields of wildlife management, fisheries, and forestry to describe the **maximum sustainable yield** that can be obtained each year from a resource (Bodmer et al. 1997). The maximum sustainable yield is the greatest amount of the resource that can be harvested each year and replaced by natural population growth. Calculations using the population growth rate and the carrying capacity (the largest population that the environment can support) are used to estimate the maximum sustainable yield. In highly controlled situations, such as plantation forestry, in which a resource can be easily quantified, it may be possible to approach the maximum sustainable yield. However, in most real-world situations, harvesting a wild species at the theoretical maximum sustainable yield is not possible, and attempts to do so often lead to an abrupt species decline (Ludwig et al. 1993; Mace and Hudson 1999).

For example, fishing industry representatives use such calculations to support their position that harvesting levels of Atlantic bluefin tuna (*Thunnus thynnus*) can be maintained at the present rate, even though the population of this species has declined by 90% in recent years (Safina 1993). In order to satisfy local business interests and protect jobs, governments often set harvesting levels too high, which results in damage to the resource base. Illegal harvesting may result in additional resource removal not accounted for in official records, such as is occurring in the whaling industry and in fishing operations in Antarctic waters. Further, a considerable proportion of the stock not harvested may be damaged as "bycatch" during harvesting operations. An additional difficulty presents itself if harvest levels are kept fairly constant even though the resource base fluctuates; a normal harvest of a fish species during a year when fish stocks are low due to unusual weather conditions may severely reduce or destroy the species.

Species that migrate across national boundaries and through international waters are particularly difficult to harvest sustainably due to the problems of coordinating international agreements and monitoring compliance. In order to protect species from total destruction, governments have begun closing fishing grounds to allow populations to recover. Such a policy, while admirable and necessary, clearly demonstrates that models of maximum sustainable yield are often inappropriate and invalid for the real world.

The hope for many overexploited species is that as they become rare, it will no longer be commercially viable to harvest them and their numbers will have a chance to recover. Unfortunately, populations of many species, such as the rhinoceros and certain wild cats, may already have been reduced so severely that they will be unable to recover. In some cases, rarity can even increase demand: As the rhinoceros becomes more rare, the price of its horn rises, making it even more valuable as a commodity on the black market. In rural areas of the developing world, desperate people may search even more intensively for the few remaining individuals of rare plant and animal species to collect and sell so that they can buy food for their families. Finding the methods to protect and manage the remaining individuals in such situations is a priority for conservation biologists.

One of the most heated debates over the harvesting of wild species has involved the hunting of whales, species to which the general public of certain Western countries has a very strong emotional attachment. After the recognition that many whale species had been hunted to dangerously low levels, the International Whaling Commission finally banned all commercial whaling in 1986. Despite that ban, certain species, such as the blue whale (*Balaenoptera musculus*) and the right whale (*Eubalaena glacialis*), which have been protected since 1967 and 1935, respectively, remain at densities far below their original numbers (Myers 1993), although the densities of other species, such as the gray whale (*Eschrichtius robustus*), appear to be recovering (Table 2.9). The slow recovery of some species may be due to illegal hunting, which continues, and to factors other than hunting that may also be responsible for unnatural whale mortality. For example, right whales frequently are killed when they collide with ships, a problem that may be occurring in other, less familiar species as well. Also, despite the ban, Japan, with its long tradition of whaling, continues to harvest limited quantities of the common small minke whale, and local fisheries in developing countries often hunt small cetaceans when there is nothing else to catch (Taylor and Dunstone 1996). Furthermore, each year thousands of dolphins and an unknown number of whales suffocate when they become entangled in deep-sea fishing

TABLE 2.9 Worldwide populations of whale species harvested by humans

Species	Numbers prior to whaling[a]	Present numbers
Baleen Whales		
Blue	200,000	9000
Bowhead	56,000	8200
Fin	475,000	123,000
Gray (Pacific stock)	23,000	21,000
Humpback	150,000	25,000
Minke	140,000	850,000
Northern right	Unknown	1300
Sei	100,000	55,000
Southern right	100,000	1500
Toothed Whales		
Beluga	Unknown	50,000
Narwhal	Unknown	35,000
Sperm	2,400,000	1,950,000

Source: After Myers 1993; Sea World 2000.

[a]Preexploitation population numbers are highly speculative.

equipment intended for tuna, cod, and other commercial fish. Efforts to require "dolphin-friendly" fishing methods have been only partly effective and have caused acrimony in trade relations between countries.

Invasive species

The geographical ranges of many species are restricted by major environmental and climatic barriers to dispersal. Mammals of North America are unable to cross the Pacific to reach Hawaii, marine fishes in the Caribbean are unable to cross Central America to reach the Pacific, and freshwater fishes in one African lake have no way of crossing the land to reach nearby, isolated lakes. Oceans, deserts, mountains, and rivers all restrict the movement of species. As a result of geographical isolation, patterns of evolution have proceeded in different ways in each major area of the world; for example, the biota of the Australia–New Guinea region, with its abundance of marsupial mammals, such as the kangaroo and koala, is strikingly different from that of the adjacent region of Southeast Asia. Islands, the most isolated of habitats, tended to evolve unique endemic biotas.

Humans have radically altered this pattern by transporting species throughout the world. In preindustrial times, people carried cultivated plants and domestic animals from place to place as they set up new farming areas and colonies. Animals such as goats and pigs were set free on uninhabited islands by European seafarers to provide food on return visits. In modern times a vast array of species have been introduced, deliberately and accidentally, into areas where they are not native (Drake et al. 1989; Vitousek et al. 1996). Many species introductions have occurred by the following means:

- *European colonization.* European settlers arriving at new colonies released hundreds of European bird and mammal species in places like New Zealand, Australia, and South Africa to make the countryside seem familiar and to provide game for hunting.

- *Horticulture and agriculture.* Large numbers of plant species have been introduced and grown in new regions as ornamentals, agricultural crops, or pasture grasses. Many of these species have escaped from cultivation and have become established in the local community.

- *Accidental transport.* Species are often transported by people unintentionally; common examples include weed seeds that are accidentally harvested with commercial seeds and sown in new localities, rats and insects that stow away aboard ships and airplanes, and disease and parasitic organisms that are transported along with their host species. Ships frequently carry exotic species in their ballast. Soil ballast dumped in port areas brings in weed seeds and soil arthropods, and water ballast introduces algae, invertebrates, and small fishes. Ballast water being released by ships into Coos Bay, Oregon, was found to contain 367 marine species originating in Japanese waters (Carlton and Geller 1993).

The great majority of **exotic species**, species occurring outside of their natural ranges due to human activity, do not become established in the places to which they are introduced because the new environment is not suitable to their needs. However, a certain percentage of species do establish themselves in their new homes, and many of these can be considered **invasive species**; that is, they increase in abundance at the expense of native species. These exotic species may displace native species through competition for limiting resources. Introduced animal species may prey upon native species to the point of extinction, or they may so alter the habitat that many natives are no longer able to persist. Invasive exotic species represent threats to 49% of the endangered species in the United States with particularly severe impacts on bird and plant species (Wilcove et al. 1998).

Many areas of the world are strongly affected by invasive species. The United States currently has more than 70 species of exotic fish,

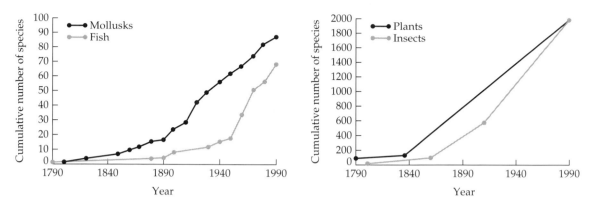

2.22 The number of exotic mollusk, fish, plant, and insect species in the United States has increased steadily over time. (After OTA 1993.)

80 species of exotic mollusks, 2000 species of exotic plants, and 2000 species of exotic insects (Figure 2.22). Many North American wetlands are completely dominated by exotic perennials: purple loosestrife (*Lythrum salicaria*) from Europe dominates marshes in eastern North America, while Japanese honeysuckle (*Lonicera japonica*) forms dense tangles in bottomlands of the southeastern United States. Insects introduced deliberately, such as European honeybees (*Apis mellifera*) and bumblebees (*Bombus* spp.), and accidentally, such as fire ants (*Solenopsis saevissima richteri*) and African honeybees (*A. mellifera adansonii* or *A. mellifera scutella*), can build up huge populations. The effects of these invasive insects on the native insect fauna can be devastating, resulting in the elimination of numerous species from the area (Porter and Savignano 1990). At some localities in the southern United States, the diversity of insect species has declined by 40% following the invasion of exotic fire ants.

Invasive species on islands. The isolation of island habitats encourages the development of a unique assemblage of endemic species, but it also leaves those species particularly vulnerable to depredations by invading species. Animals introduced onto islands have efficiently preyed upon endemic animal species and have grazed down native plant species to the point of extinction. Introduced plant species with tough, unpalatable foliage are better able to coexist with the introduced grazers, such as goats and cattle, than are the more palatable native plants, so the exotics begin to dominate the landscape as the native vegetation dwindles. Island animal species adapted to a community with few mammalian predators may have few defenses against introduced predators, such as cats and rats. Moreover, island species often have no natural immunities to mainland diseases; when exotic species are introduced to the island, they frequently carry pathogens or parasites that, though relatively harmless to the carrier, can devastate the native populations.

Two examples illustrate the effects of introduced species on the biota of islands:

- *Plants of Santa Catalina Island*. Forty-eight native plant species have been eliminated from Santa Catalina Island off the coast of California, primarily due to grazing by introduced goats and other exotic mammals. Removal of goats from part of the island has led to the reappearance of several native plant species.

- *Birds of the Pacific Islands*. The brown tree snake (*Boiga irregularis*; Figure 2.23) has been accidentally introduced onto a number of islands in the Pacific Ocean where it is devastating bird populations. The snake eats eggs, nestlings, and adult birds; on Guam alone, it has reduced 10 endemic bird species to the point of extinction. Visitors have remarked on the absence of birdsong (Jaffe 1994).

Invasive species in aquatic habitats. Exotic species can have severe effects on lakes, streams, and even entire marine ecosystems. Freshwater communities in particular are similar to oceanic islands in that they are isolated habitats surrounded by vast stretches of uninhabitable terrain, and are likewise vulnerable to the introduction of exotic species. Commercial and sport fish species are often introduced into aquatic environments where they do not naturally occur.

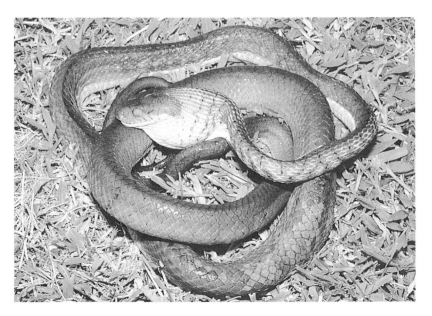

2.23 The brown tree snake (*Boiga irregularis*) has been introduced onto many Pacific islands, where it has devastated populations of endemic birds. This adult snake has just swallowed a bird. (Photograph by Julie Savidge.)

Over 120 fish species have been introduced into marine and estuar-ine systems and inland seas; although some of these introductions have been deliberate attempts to increase fisheries, most of them were the unintentional result of canal building and the transport of ballast water in ships (Baltz 1991). Often these exotic species are larger and more aggressive than the native fish fauna, and through a combination of competition and direct predation, they eventually drive the local fishes to extinction.

Aggressive aquatic exotics include plants and invertebrate ani-mals as well as fishes. One of the most alarming recent invasions in North America was the arrival in 1988 of the zebra mussel (*Dreissena polymorpha*) in the Great Lakes. This small, striped native of the Caspian Sea apparently was a stowaway in the ballast of a European tanker. Within two years zebra mussels had reached densities of 700,000 individuals per square meter in parts of Lake Erie, choking out native mussel species in the process (Stolzenburg 1992). Zebra mussels have been found in the Detroit, Cumberland, and Tennessee Rivers. As it spreads southward, this exotic species is causing enor-mous economic damage to fisheries, dams, power plants, and boats, as well as devastating the aquatic communities it encounters.

The ability of exotic species to invade. Why are some exotic spe-cies so easily able to invade and dominate new habitats and dis-place native species? One reason is the absence of their natural predators, pests, and parasites in the new habitat. Rabbits intro-duced into Australia, for example, spread uncontrollably, grazing native plants to the point of extinction, because there were no effec-tive checks on their numbers. Control efforts have focused in part on introducing into Australia diseases that help control rabbit pop-ulations elsewhere.

Human activity may create unusual environmental conditions, such as increased levels of soil disturbance, an increased incidence of fire, or enhanced light availability, to which exotic species can adapt more readily than native species. The highest concentrations of inva-sive exotics are often found in habitats that have been most altered by human activity. In Southeast Asia, for example, progressive de-gradation of forests results in a progressively smaller proportion of native species living in the habitat (Figure 2.24). As people in-creasingly alter the environment through air and water pollution, agriculture, logging, fishing, and global climate change, the problem that invasive species pose will continue to increase.

Invasive species are considered to be the most serious threat fac-ing the biota of the United States National Park system. While the effects of habitat degradation, fragmentation, and pollution can potentially be corrected and reversed in a matter of years or de-

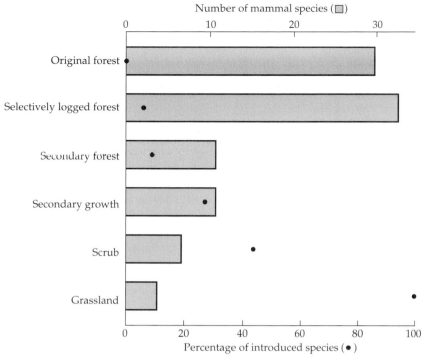

2.24 Progressive degradation of Southeast Asian forests by logging and farming not only decreases the number of species of nonflying native mammals, but increases the percentage of introduced species. Only introduced rats are present in the final grassland stage. (From Harrison 1968.)

cades as long as the original species are present, exotic species that are well established may be impossible to remove from communities. Invasive exotic species may have built up such large numbers and become so widely dispersed and so thoroughly integrated into the community that eliminating them may be extraordinarily difficult and expensive. Widespread native trees all across North America are being attacked and even destroyed by introduced insects and fungi, causing economic and ecosystem damage on a continent-wide scale.

An additional class of invasives is species that have increased their ranges within continental areas because they are suited to the ways in which humans have altered the environment. Within North America, fragmentation of forests, suburban development, and easy access to garbage have allowed the numbers and ranges of coyotes, red foxes, and herring gulls to increase. As these aggressive species increase, they do so at the expense of native species that are less competitive and less able to resist predation. Native species that build up to unusually high numbers because they can adapt well to human activities represent a further challenge to the management of vulnerable species and protected areas.

Another special class of invasives is those that have close relatives in the native biota. When these exotic species hybridize with native species and varieties, unique genotypes may be eliminated from local populations, and taxonomic boundaries may become obscured. Such appears to be the fate of native trout species when confronted by commercial species. In the American Southwest, the Apache trout (*Oncorhynchus apache*) has had its range reduced by habitat destruction and by competition with introduced fish species. The species has also hybridized extensively with the rainbow trout (*O. mykiss*), an introduced sport fish (Dowling and Childs 1992).

Disease

Infections by disease organisms are common in both wild and captive populations (Aguirre and Starkey 1994; McCallum and Dobson 1995). Infections may come from **microparasites**, such as viruses, bacteria, fungi, and protozoa, or **macroparasites**, such as helminth worms and parasitic arthropods. Such diseases may be the single greatest threat to some rare species. The last population of black-footed ferrets (*Mustela nigrepes*) known to exist in the wild was destroyed by the canine distemper virus in 1987 (Miller et al. 1996). One of the main challenges of managing the captive breeding program for black-footed ferrets has been protecting the captives from canine distemper, human viruses, and other diseases; this is being done through rigorous quarantine measures and subdivision of the captive colony into geographically separated groups. The black-footed ferret remains extremely susceptible to canin distemper, which is present in carnivore populations throughout the ferret's potential range. As a result, animals released into the wild in 1991 remain vulnerable to an epidemic.

Three basic principles of epidemiology have obvious practical implications for the captive breeding and management of rare species. First, both captive and wild animals in dense populations may face increased direct pressure from parasites and diseases. In fragmented conservation areas, populations of animals may temporarily build up to unnaturally high densities that promote high rates of disease transmission. In natural situations, the level of infection is typically reduced when animals migrate away from their droppings, saliva, old skin, and other sources of infection. In unnaturally confined situations, the animals remain in contact with these potential sources of infection, and disease transmission increases. In zoos, colonies of animals are often caged together in a small area. Consequently, if one animal becomes infected, the parasite can spread rapidly through the population.

Second, the indirect effects of habitat destruction can increase an organism's susceptibility to disease. When a host population is crowded into a smaller area because of habitat destruction, there will often be a deterioration in habitat quality and food availability, leading to lowered nutritional status, weaker animals, and a greater susceptibility to infection. Crowding can also lead to social stress within a population, which lowers the animals' resistance to disease. Pollution may make individuals more susceptible to infection by pathogens, particularly in the aquatic environment.

Third, in many conservation areas, zoos, national parks, and new agricultural areas, species come into contact with other species that they would rarely or never encounter in the wild—including humans and domestic animals—so that infections can spread from one species to another (Figure 2.25). Certain emerging infectious diseases, such as human immunodeficiency virus (HIV) and Ebola virus, appear to have even spread from wildlife populations to both humans and domestic animals. Once infected with exotic diseases, captive animals cannot be returned to the wild without threatening the entire wild population with infection. Also, a species that is common and fairly resistant to a disease can act as a reservoir for the disease, which can then infect populations of highly susceptible species.

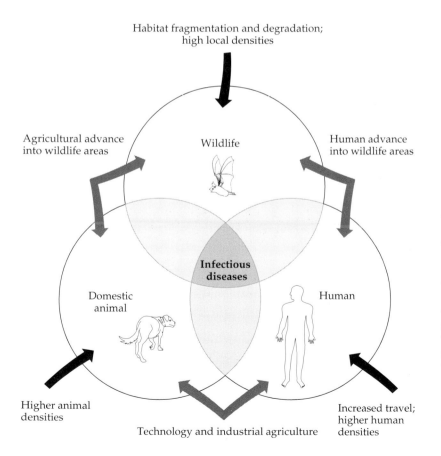

Habitat fragmentation and degradation; high local densities

Agricultural advance into wildlife areas

Wildlife

Human advance into wildlife areas

Infectious diseases

Domestic animal

Human

Higher animal densities

Technology and industrial agriculture

Increased travel; higher human densities

2.25 Infectious diseases, such as rabies, lyme disease, influenza, hantavirus, and canine distemper, spread among wildlife populations, domestic animals, and humans as a result of increasing population densities and the advance of agriculture and human settlements into wildlife areas. The shaded areas of overlap indicate diseases shared between the three groups. Black arrows indicate factors contributing to higher rates of infection; shaded arrows indicate factors contributing to the spread of disease among the three groups. The figure illustrates the example of rabies—bats, dogs, and humans are all susceptible to the rabies virus. (After Daszak et al. 2000.)

2.26 Populations of flowering dogwood (*Cornus florida*) are currently declining in eastern North American forests because of anthracnose disease caused by the introduced fungus *Discula destructiva*. (Photograph by Jonathan P. Evans.)

For example, apparently healthy African elephants can transmit a fatal herpes virus to related Asian elephants when kept together in zoos (Richman et al. 1999). During the early 1990's in Tanzania's Serengeti National Park, about 25% of the lions were killed by canine distemper, apparently contracted from one or more of the 30,000 domestic dogs living near the park (Morell 1994). Diseases can decimate even common species: North American chestnut trees (*Castanea dentata*), once common throughout the eastern United States, have been virtually obliterated in that region by an ascomycete fungus carried by Chinese chestnut trees imported to New York City. An introduced fungus is now killing flowering dogwood (*Cornus florida*) in much of its native range (Figure 2.26).

Vulnerability to Extinction

When environments are damaged by human activity, the population sizes of many species will be reduced, and some species will become extinct. Ecologists have observed that not all species have an equal probability of going extinct; particular categories of species are especially vulnerable to extinction, and need to be carefully monitored and managed in conservation efforts (Terborgh 1974; Pimm et al. 1988; Gittleman 1994).

- *Species with narrow geographical ranges*. Some species occur at only one or a few sites in a restricted geographical range, and if that whole range is affected by human activity, then the species may become extinct. Bird species on oceanic islands provide many examples of species with restricted ranges that have become extinct; many fish species confined to a single lake or a single watershed have also disappeared (Figure 2.27).

- *Species with only one or a few populations.* Any one population of a species may become locally extinct as a result of environmental factors such as earthquakes, fires, or outbreaks of disease, as well as the impact of human activity. Species with many populations are therefore less vulnerable to global extinction than are species that consist of only one or a few populations.

- *Species in which population size is small, sometimes called "the small population paradigm"* (Caughley and Gunn 1996). Small populations are more likely to become locally extinct than large populations due to their greater vulnerability to demographic and environmental variation and loss of genetic variability, as described in Chapter 3. Species that characteristically have small population sizes, such as large predators and extreme specialists, are more likely to become extinct than species that typically have large populations (Mace 1994).

- *Species in which population size is declining, sometimes called the "declining population paradigm."* Population trends tend to continue, so a population showing signs of decline is likely to become extinct unless the cause of decline is identified and corrected.

- *Species with low population densities.* A species with a low population density—few individuals per unit of area—is likely to have

2.27 Species of desert pupfish (*Cyprinodon* spp.) of the southwestern United States are highly endangered by the degradation and disappearance of their very restricted habitat—saline desert ponds. Some species are found in only one or a few ponds. (Photograph © Ken Kelley, Zoological Society of San Diego.)

only small populations remaining in each fragment if its range is broken up by human activities. Within each fragment the population size may be too small to allow the species to persist, and it will gradually die out across the landscape.

- *Species that need large home ranges.* Species in which individual animals or social groups need to forage over a wide area are prone to die off when part of their range is damaged or fragmented by human activity.

- *Animal species with large body sizes.* Large animals tend to have larger individual ranges, require more food, and are more easily hunted to extinction by humans than are small animals. Top carnivores are often killed because they compete with humans for wild game, sometimes prey on domestic animals or people, and are hunted for sport. Within guilds of species, the largest species—the largest carnivore, the largest lemur, the largest whale—is often the most prone to extinction. In Sri Lanka, for example, the largest species of carnivores (leopards and eagles) and the largest species of herbivores (elephants and deer) are currently at the greatest risk of extinction (Erdelen 1988).

- *Species that are not effective dispersers.* In the natural world, environmental changes prompt species to adapt, either behaviorally or physiologically, to the new conditions of their habitat. Species unable to adapt to changing environments must either migrate to more suitable habitat or face extinction. The rapid pace of human-induced changes often prevents adaptation, leaving migration as the only alternative. Species that are unable to cross the roads, farmlands, and other disturbed habitats created by human activity are doomed to extinction as their original habitat is altered by pollution, invasive species, and global climate change. Among North American aquatic invertebrates, poor dispersal activity helps to explain why 68% of the snail and mussel species are extinct or in danger of extinction in contrast with only 20% of the dragonfly species, which can fly between the aquatic sites needed by their larval stages (Stein and Flack 1997).

- *Seasonal migrants.* Species that migrate seasonally depend on two or more distinct habitat types. If either habitat is damaged, then the species may be unable to persist. The billion songbirds of 120 species that migrate each year between Canada and the neotropics depend on suitable habitat in both locations to survive and breed. Also, if barriers to dispersal are created between the needed habitats by roads, fences, or dams, a species may be unable to complete its life cycle. Salmon species that are prevented by dams from swimming up rivers to spawn are a striking example of this problem.

- *Species with little genetic variability.* Genetic variability within a population can sometimes allow a species to adapt to a changing environment. Species with little or no genetic variability may have a greater tendency to become extinct when a new disease, a new predator, or some other change occurs in the environment. For example, extremely low genetic variability is considered to be a contributing factor to lack of disease resistance in the cheetah (*Acinonyx jubatus*) (O'Brien and Evermann 1988), although environmental factors may be the predominant reason for the decline of this species (Caro and Laurenson 1994).

- *Species with specialized niche requirements.* Some species are confined to a single unusual habitat type that is scattered and rare, such as limestone outcrops or caves. If the habitat is damaged by human activity, then species may not be able to survive. Species with highly specific dietary requirements are also at risk. At the extreme, for instance, are species of mites that feed only on the feathers of a single bird species. If the bird species goes extinct, so do its associated feather mite species.

- *Species that are characteristically found in stable environments.* Many species are adapted to environments where disturbance is minimal, such as old stands of tropical rain forest. Such species often have slow growth, low rates of reproduction, and few offspring. When these forests are logged, grazed, burned, or otherwise altered by human activity, many native species are unable to tolerate the changed microclimatic conditions (more light, less moisture, greater temperature variation) and competition with early successional species and invasive species.

- *Species that form permanent or temporary aggregations.* Species that group together in specific places are highly vulnerable to local extinction. For example, bats forage widely at night but typically roost together in particular caves during the day. Hunters entering those caves during the day can rapidly harvest every individual in the population. Herds of bison, flocks of passenger pigeons, and schools of fish all represent aggregations that have been exploited and completely harvested by people (to the point of extinction, in the case of the passenger pigeon). Some species of social animals may be unable to persist when their population size falls below a certain number because they can no longer forage, mate, or defend themselves.

- *Species that are hunted or harvested by people.* Utility has often been the prelude to extinction. Overharvesting can rapidly reduce the population size of a species that is economically valuable to humans. If hunting and harvesting are not regulated, either by law or by local custom, the species can be driven to extinction.

These characteristics of extinction-prone species are not independent, but tend to group together into categories of characteristics. For example, species with large body sizes tend to have low population densities and large home ranges—all characteristics of extinction-prone species. By identifying characteristics of extinction-prone species, conservation biologists can anticipate which species will need to be protected and managed.

Summary

1. Human activity has already driven many species to extinction. Since 1600, about 2.1% of the world's mammal species and 1.3% of its bird species have gone extinct. The rate of extinction is accelerating, and many extant species are teetering on the brink of extinction. More than 99% of modern species extinctions are attributable to human activity.

2. Many species that occupy islands are especially vulnerable to extinction because they are endemic to only one or a few islands. An island biogeography model has been used to predict that current rates of habitat destruction will result in the loss of about 25,000 species per year over the next 10 years. Many biological communities are being gradually impoverished by local extinctions of species.

3. Slowing human population growth is part of the solution to the biological diversity crisis. In addition, large-scale industrial activities, logging, and agriculture are often unnecessarily destructive to the natural environment in their pursuit of short-term profit. Efforts to reduce the high consumption of natural resources in wealthy industrialized countries and to eliminate poverty in developing countries are also important parts of the overall strategy to protect biological diversity.

4. The major threat to biological diversity is loss of habitat, and the most important means of protecting biological diversity is preserving habitat. Habitats particularly threatened with destruction are rain forests, tropical dry forests, wetlands in all climates, temperate grasslands, mangrove forests, and coral reefs.

5. Habitat fragmentation is the process whereby a large, continuous area of habitat is both reduced in area and divided into two or more fragments. Habitat fragmentation can lead to the rapid loss of the species remaining because it creates barriers to the normal processes of dispersal, colonization, and foraging. Environmental conditions in the fragments may change, and pests may become more common.

6. Environmental pollution eliminates many species from biological communities even where the structure of the community is

not obviously disturbed. The range of environmental pollution includes excessive use of pesticides; contamination of water sources with industrial wastes, sewage, and fertilizers; and air pollution resulting in acid rain, excess nitrogen deposition, photochemical smog, and ozone.

7. Global climate change, in particular warmer weather, is already occurring because of the large amounts of carbon dioxide and other greenhouse gases entering the atmosphere, produced by the burning of fossil fuels. Predicted temperature changes could be so rapid that many species will be unable to adjust their ranges and will probably become extinct.

8. Growing rural poverty, increasingly efficient methods of hunting and harvesting, and the globalization of the economy combine to encourage the overexploitation of many species to the point of extinction. Traditional societies had customs to prevent overharvesting of resources, but these customs are being abandoned as societies modernize.

9. Humans have deliberately and accidentally moved thousands of species to new regions of the world. Some of these exotic species have become invasive, greatly increasing in abundance and outcompeting or eating native species.

10. Levels of disease and parasites often increase when animals are confined to a nature reserve and cannot disperse over a wide area. Animals held in captivity and living in degraded environments are particularly prone to high levels of disease. Diseases sometimes spread between related species of animals.

11. Species most vulnerable to extinction have particular characteristics, such as very narrow geographic ranges, only one or a few populations, small population size, declining population size, and economic value to humans, that lead to overexploitation.

Suggested Readings

Birkeland, C. (ed.). 1997. *The Life and Death of Coral Reefs*. Chapman and Hall, New York. Extensive background on this important ecosystem, including discussion of the threats that confront it.

Carson, R. 1962. *Silent Spring*. Reprinted in 1982 by Penguin, Harmondsworth, England. This book's descriptions of the harmful effects of pesticides on birds created heightened public awareness of environmental problems when it was first published.

Daszak, P., A. A. Cunningham and A. D. Hyatt. 2000. Emerging infectious diseases of wildlife—threats to biodiversity and human health. *Science* 287:443–449. Review of new diseases that spread among wildlife, humans, and domestic animals.

Gates, D. M. 1993. *Climate Change and Its Biological Consequences*. Sinauer Associates, Sunderland, MA. Clear and thorough description of both past and current climate changes and their effects.

Hardin, G. 1993. *Living within Limits: Ecology, Economics, and Population Taboos*. Oxford University Press, New York. Blunt advice on controlling human numbers.

Laurance, W. F. and R. O. Bierregaard, Jr. (eds.). 1997. *Tropical Forest Remnants: Ecology, Management and Conservation of Fragmented Communities*. The University of Chicago Press, Chicago. Comprehensive treatment of habitat fragmentation.

Ludwig, D., R. Hilborn and C. Walters. 1993. Uncertainty, resource exploitation, and conservation: Lessons from history. *Science* 260: 17, 36. Excellent short statement about why commercial exploitation so often destroys its resource base. See the 1993 volume of the journal *Ecological Applications* for more articles on this topic.

MacArthur, R. H. and E. O. Wilson. 1967. *The Theory of Island Biogeography*. Princeton University Press, Princeton, NJ. This classic text outlining the island biogeography model has been highly influential in shaping modern conservation biology.

Quammen, D. 1996. *The Song of the Dodo: Island Biogeography in an Age of Extinctions*. Scribner, New York. Popular account of early and modern explorations, and of island biogeography theory.

Rochelle, J. A., L. A. Lehman and J. Wisniewski (eds.). 1999. *Forest Fragmentation: Wildlife and Management Implications*. Koninkliijke Brill NV, Leiden, Netherlands. Various alternative forest management practices affect animal populations.

Schneider, S. 1998. *Laboratory Earth: The Planetary Gamble We Can't Afford to Lose*. Basic Books, New York. Leading authority clearly explains the complex ideas of global climate change and why it is vitally important to take action.

Stearns, B. P. and S. C. Stearns. 1999. *Watching, From the Edge of Extinction*. Yale University Press, New Haven, CT. Captures the drama, excitement, and frustrations of conservation biologists involved in last-ditch efforts to save species.

Taylor, V. J. and N. Dunstone (eds.). 1996. *The Exploitation of Mammal Populations*. Chapman and Hall, London. Experts attempt to determine if sustainable harvesting is possible. Includes excellent case studies.

Terborgh, J. 1999. *Requiem for Nature*. Island Press, Washington D.C. The multiple threats faced by biological diversity must be considered realistically and without illusions.

Union of Concerned Scientists. 1999. *Global Warming: Early Signs*. Cambridge, MA. Summaries of evidence for global warming, and its effects.

Vitousek, P. M., C. M. D'Antonio, L. L. Loope and R. Westerbrooks. 1996. Biological invasions as global environmental change. *American Scientist* 84: 468–478. Excellent review article describing impact of exotic species on ecosystems, human health, and the economy.

Watling, L. and E. Norse. 1998. Disturbance of the seabed by mobile fishing gear: A comparison to forest clearcutting. *Conservation Biology* 12: 1180–1197. Special section devoted to fishing impacts.

Wilcove, D. S. 1999. *The Condor's Shadow: The Loss and Recovery of Wildlife in America*. W. H. Freeman, New York. A senior ecologist with Environmental Defense reviews human impacts on wildlife, offering his opinions on what conservation methods offer some hope.

Chapter 3

Conservation at the Population and Species Levels

Conservation efforts often seek to protect species that are declining in number and in danger of becoming extinct. As we have already discussed, the activities of humans often impose serious restrictions on species. In order to successfully maintain species under these restrictions, conservation biologists must determine the stability of populations under certain circumstances. Is a species declining, and does it require special attention to prevent it from going extinct? Alternatively, will a population of an endangered species persist or even increase in a nature reserve?

Many national parks and wildlife sanctuaries have been created to protect "charismatic megafauna," such as lions, tigers, and bears, which are important as national symbols and as tourist attractions. However, merely designating the communities in which these species live as protected areas may not stop their decline and extinction, even when they are legally protected. Sanctuaries generally are created only after most populations of a threatened species have already been severely reduced by habitat loss, habitat degradation, habitat fragmentation, or overharvesting. Under these circumstances, a species tends to dwindle rapidly toward extinction. Also, individuals outside the reserve boundaries remain unprotected and at risk. To design a conservation plan that will protect an endangered species—and, hopefully, restore it—it is vitally important to understand the status of its populations in the wild, how those populations react under different conditions, and the species' natural history and ecology. This chapter examines how conservation biologists use a population-level approach to protect and restore species.

Conserving Species by Conserving Populations

As a general rule, an adequate conservation plan for an endangered species requires that as many individuals as possible be preserved within the greatest possible area of protected habitat. However, this general statement does not provide specific guidelines to assist planners, land managers, politicians, and wildlife biologists who are trying to protect species from extinction. The problem is exacerbated by the fact that planners often must proceed without a firm understanding of the range and habitat requirements of the species. For example, to preserve the red-cockaded woodpecker, does longleaf pine habitat in the southeastern United States need to be preserved for 50, 500, 5000, 50,000, or even more individuals? Furthermore, planners must reconcile conflicting demands on finite resources—a problem vividly demonstrated throughout the world by efforts to balance the powerful economic forces promoting coastal development with the need to protect endangered marine and coastal habitats.

In a groundbreaking paper, Shaffer (1981) defined the number of individuals necessary to ensure the survival of a species as its **minimum viable population** (**MVP**): "A minimum viable population for any given species in any given habitat is the smallest isolated population having a 99% chance of remaining extant for 1000 years despite the foreseeable effects of demographic, environmental, and genetic stochasticity, and natural catastrophes." In other words, an MVP is the smallest population that can be predicted to have a very high chance of persisting for the foreseeable future. Shaffer emphasized the tentative nature of this definition, stating that the survival probabilities could be set at 95%, 99%, or any other percentage, and that the time frame might similarly be adjusted, for example, to 100 years or 500 years. The key point of the MVP is that it allows a quantitative estimate to be made of how many individuals are needed to preserve a species (Menges 1991).

Shaffer (1981) compares MVP protection efforts to flood control efforts. In planning flood control systems and regulating building on wetlands, it is not sufficient to use average annual rainfall as a guideline. We recognize the need to plan for severe flooding, which may occur only once every 50 years. Likewise, in protecting natural systems, we understand that certain catastrophic events, such as massive hurricanes, earthquakes, forest fires, volcanic eruptions, epidemics, and die-offs of food items, may occur at even longer intervals. To plan for the long-term protection of an endangered species, we not only have to provide for the requirements of the species in average years, but also for its needs in exceptional years. In drought years, for instance, animals may migrate well beyond their normal ranges to obtain the water they need to survive.

Obtaining an accurate estimate of the MVP for a particular species may require a detailed demographic study of the population and an environmental analysis of the site, which may be expensive and require months or even years of research (Thomas 1990). Some biologists have suggested a general rule of attempting to protect 500–5000 individuals for vertebrate species, as this number seems adequate to preserve genetic variability (Lande 1988, 1995). Protecting this number of individuals may be adequate to allow a minimum number of individuals to survive in catastrophic years and return the population to former levels. For species with extremely variable population sizes, such as certain invertebrates and annual plants, protecting a population of about 10,000 individuals might be an effective strategy.

Once a minimum viable population size has been established for a species, the **minimum dynamic area** (**MDA**), the amount of suitable habitat necessary for maintaining the MVP, can be estimated. The MDA can be estimated by studying the home range sizes of individuals and groups. Estimates have been made that reserves of 10,000 to 100,000 ha are needed to maintain many small mammal populations (Schonewald-Cox 1983). To preserve populations of wide-ranging grizzly bears in Canada, for instance, the areas needed are enormous: 49,000 km^2 for 50 individuals and 2,420,000 km^2 for 1000 individuals (Noss and Cooperrider 1994).

Small Populations Are Especially Threatened

One of the best-documented examples of determining minimum viable population size comes from a study of the persistence of 120 bighorn sheep (*Ovis canadensis*) populations in the deserts of the southwestern United States (Berger 1990, 1999). Some of these populations have been followed for up to 70 years. The study made the striking observation that 100% of the unmanaged populations with fewer than 50 individuals went extinct within 50 years, while virtually all of the populations with more than 100 individuals persisted

3.1 The relationship between initial population size (N) of bighorn sheep and the percentage of populations that persist over time. Almost all populations with more than 100 sheep persisted beyond 50 years, while populations with fewer than 50 sheep died out within 50 years. Not included are small populations that were actively managed and augmented by the release of additional animals. (After Berger 1990.)

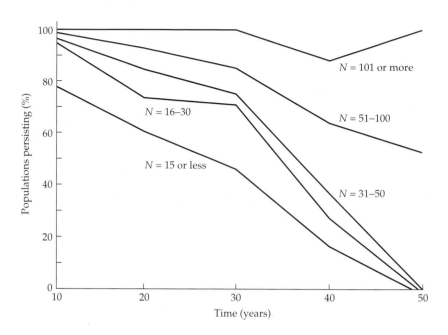

over this time period (Figure 3.1). Despite the factors hindering the survival of small populations, habitat management by government agencies and the release of additional animals have allowed some small populations to persist that might have gone extinct.

Field evidence from long-term studies of birds on the California Channel Islands supports the need for large populations to ensure population persistence: only the populations of more than 100 pairs had a greater than 90% chance of surviving for 80 years (Jones and Diamond 1976). On the other hand, there is no need to give up entirely on small populations: Many populations of birds have apparently survived for 80 years with 10 or fewer breeding pairs.

The Problems of Small Populations

Exceptions notwithstanding, large populations are needed to protect most species, and species with small populations are in real danger of going extinct. Small populations are subject to rapid decline in numbers and local extinction for three main reasons: (1) genetic problems resulting from loss of genetic variability, inbreeding, and genetic drift; (2) demographic fluctuations due to random variations in birth and death rates; and (3) environmental fluctuations due to variations in predation, competition, incidence of disease, and food supply, as well as natural catastrophes resulting from single events that occur at irregular intervals, such as fires, floods, or droughts.

Loss of genetic variability

Genetic variability is important in allowing populations to adapt to a changing environment (see Chapter 1). Individuals with certain alleles or combinations of alleles may have just the characteristics needed to survive and reproduce under the new conditions. Within a population, particular alleles may vary in frequency from common to very rare. In small populations, allele frequencies may change from one generation to the next simply due to chance, depending on which individuals mate and leave offspring, a process known as **genetic drift**. When an allele is at a low frequency in a small population, it has a significant probability of being lost in each generation due to chance. Considering the theoretical case of an isolated population in which there are two alleles per gene, Wright (1931) proposed a formula to express the proportion of original heterozygosity (individuals possessing two different allele forms of the same gene) remaining after each generation (H) for a population of breeding adults (N_e):

$$H = 1 - \frac{1}{2N_e}$$

According to this equation, a population of 50 individuals would have 99% of its original heterozygosity after one generation due to the loss of rare alleles, and still have 90% after 10 generations. However, a population of 10 individuals would have only 95% of its original heterozygosity after 1 generation and only 60% after 10 generations (Figure 3.2).

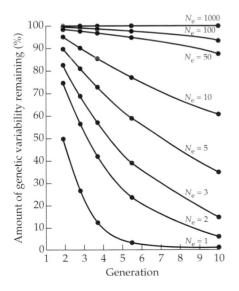

3.2 Genetic variability is lost randomly over time through genetic drift. This graph shows the average percentage of genetic variability remaining over 10 generations in theoretical populations of various effective population sizes (N_e). After 10 generations, there is a loss of genetic variability of approximately 40% with a population size of 10, 65% with a population size of 5, and 95% with a population size of 2. (From Meffe and Carroll 1997.)

This formula demonstrates that significant losses of genetic variability can occur in isolated small populations, particularly those on islands and in fragmented landscapes. However, migration of individuals among populations and the regular mutation of genes tend to increase the amount of genetic variability within a population and balance the effects of genetic drift. Even a low frequency of movement of individuals between populations minimizes the loss of genetic variability associated with small population size (Mills and Allendorf 1996; Bryant et al. 1998). If only one new immigrant arrives every generation in an isolated population of about 100 individuals, the effect of genetic drift will be negligible. Such **gene flow** appears to be the major factor preventing the loss of genetic variability in small populations of Galápagos finches (Grant and Grant 1992). Although the mutation rates found in nature—about 1 in 1000 to 1 in 10,000 per gene per generation—may make up for random losses of alleles in large populations, they are ineffective at countering genetic drift in small populations of about 100 individuals or fewer.

In addition to theories, lab experiments, and simulations, field data also show that small population size leads to a more rapid loss of alleles from the population (Frankham 1996). In a New Zealand conifer species, populations of less than 1000 individuals suffered much higher losses of genetic variability than populations of more than 10,000 individuals (Billington 1991). An extensive review of studies of genetic variability in plants showed that only 8 of 113 species had no measurable genetic variability and that most of those 8 species had very limited ranges (Hamrick and Godt 1989).

Small populations subjected to genetic drift have greater susceptibility to a number of deleterious genetic effects, such as inbreeding depression, outbreeding depression, and loss of evolutionary flexibility. These factors may contribute to a decline in population size and a greater probability of extinction (Thornhill 1993; Loeschcke et al. 1994; Avise and Hamrick 1996).

Inbreeding depression. A variety of mechanisms prevents inbreeding in most natural populations. In large populations of most animal species, individuals do not normally mate with close relatives. Individuals often disperse away from their place of birth, or are inhibited from mating with relatives through unique individual odors or other sensory cues. In many plants, a variety of morphological and physiological mechanisms encourage cross-pollination and prevent self-pollination. In some cases, however, particularly when population size is small and no other mates are available, these mechanisms fail to prevent inbreeding. Mating among close relatives, such as parents and their offspring, siblings, and cousins, and self-fertilization in hermaphroditic species may result in **inbreeding depression**, a condition characterized by fewer offspring or offspring that are weak or sterile (Ralls et al. 1988). For example, plants of the scarlet gilia (*Ipomopsis aggregata*) that come from populations with fewer than 100 individuals produce smaller seeds with a lower rate of seed germination and exhibit a greater susceptibility to environmental stress than do plants from larger populations (Figure 3.3). Such symptoms, associated with inbreeding depression and loss of genetic variation, are lessened when plants from small populations are cross-pollinated using pollen from plants of large populations. In Illinois, isolated small populations of prairie chickens (*Tympanuchus cupido pinnatus*) were showing the effects of declining genetic variation, with lowered fertility and lowered rates of egg-hatching (Westemeier et al. 1998). When individuals from large, genetically diverse populations were released into the populations, egg viability was restored, demonstrating the importance of maintaining genetic variation.

The most plausible explanation for inbreeding depression is that it allows the expression of harmful alleles inherited from both parents (Barrett and Kohn 1991). Inbreeding depression can be a severe problem in small captive populations in zoos and domestic breeding programs.

Outbreeding depression. Individuals from geographically separate and genetically distinct populations rarely mate in the wild, due not only to physical separation but also to a number of behavioral, physiological, and morphological mechanisms that ensure that mating takes place only between genetically similar individuals from

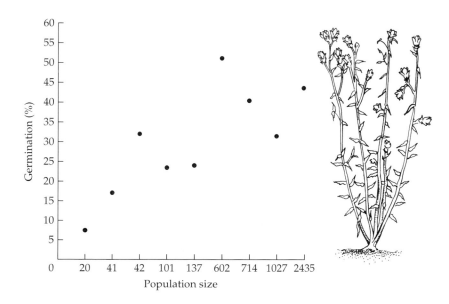

3.3 Seed germination in populations of the scarlet gilia (*Ipomopsis aggregata*) from montane Arizona is lower in small populations (fewer than 150 individuals) in comparison to larger populations. Seed germination is strongly reduced in the smallest populations. (After Heschel and Paige 1995.)

the same species. However, when a species is rare or its habitat is damaged, outbreeding—mating between separate populations—may occur. Individuals unable to find mates within their own population may mate with individuals from other populations. The resulting offspring may be weak or sterile due to a lack of compatibility of the chromosomes and enzyme systems inherited from their different parents, a condition known as **outbreeding depression** (Templeton 1986; Thornhill 1993). These hybrid offspring may also no longer have the precise combination of genes that allowed individuals to survive in a particular set of local conditions. At the extreme, outbreeding depression can also result from matings between individuals of closely related species. To guard against outbreeding depression, captive breeding programs must avoid the pairing of individuals from closely related species and from the extremes of the species' geographical range. When experimental populations of rare species are created using individuals from separate populations, they need to be monitored for the possible effects of outbreeding depression.

Outbreeding depression may be particularly important in plants, in which mate selection is to some degree a matter of the chance movement of pollen. A rare plant species growing near a closely related common species may become overwhelmed by the pollen of the common species, leading to sterile offspring or a blurring of species boundaries (Ellstrand 1992).

Loss of evolutionary flexibility. Rare alleles and unusual combinations of alleles that confer no immediate advantages may be uniquely suited for some future set of environmental conditions. Loss of genetic variability in a small population may limit the ability of the population to respond to long-term changes in the environment, such as pollution, new diseases, or global climate change (Falk and Holsinger 1991). Without sufficient genetic variability, a species may become extinct.

Effective population size

How many individuals are needed to maintain genetic variability in a population? Franklin (1980) provided evidence that 50 individuals might be the minimum number necessary to maintain genetic variability. This figure was based on the practical experience of animal breeders, which indicates that animal stocks can be maintained with a loss of 2%–3% of the variability per generation. Wright's formula shows that a population of 50 individuals will lose only 1% of its variability per generation, so using this figure would be erring on the safe side. However, because Franklin's estimate is based on work with domestic animals, its applicability to the wide range of wild species is uncertain. Using data on mutation rates in *Drosophila* fruit flies, Franklin (1980) suggested that in populations of 500 individuals, the rate of new genetic variability arising through mutation might balance the variability being lost due to small population size. This range of values has been referred to as the 50/500 rule: Isolated populations need to have at least 50 individuals, and preferably 500 individuals to maintain genetic variability.

The 50/500 rule is difficult to apply in practice because it assumes that a population is composed of N individuals that all have an equal probability of mating and having offspring. However, many individuals in a population do not produce offspring due to factors such as age, poor health, sterility, malnutrition, small body size, or social structures that prevent some animals from finding mates. As a result of these factors, the **effective population size** (N_e) of breeding individuals is often substantially smaller than the actual population size. Because the rate of loss of genetic variability is based on the effective population size, loss of genetic variability can be quite severe even when the actual population size is much larger (Kimura and Crow 1963; Nunney and Elam 1994). An effective population size that is smaller than expected can exist under any of the following circumstances.

Unequal sex ratio. By random chance, the population may consist of unequal numbers of males and females. If, for example, a population of a monogamous goose species (in which one male and one female form a long-lasting pair bond) consists of 20 males and 6 females, only 12 individuals will be involved in mating activity. In this case, the effective population size is 12, not 26. In other animal species, social systems may prevent many individuals from mating even though they are physiologically capable of doing so: In elephant seals, for example, a dominant male may control a large group of females and prevent other males from mating with them.

The effect of unequal numbers of breeding males and females on N_e can be described by the general formula:

$$N_e = \frac{4N_m N_f}{N_m + N_f}$$

where N_m and N_f are the numbers of breeding males and breeding females, respectively, in the population. In general, as the sex ratio of breeding individuals becomes increasingly unequal, the ratio of the effective population size to the number of breeding individuals (N_e/N) also goes down.

Variation in reproductive output. In many species, the number of offspring varies substantially among individuals. This is particularly true in plants, in which some individuals may produce a few seeds while others produce thousands of seeds. Unequal production of offspring leads to a substantial reduction in N_e of up to 85% in many species because a few individuals in the present generation will be disproportionately represented in the gene pool of the next generation.

Population fluctuations. In some species, population size varies dramatically from generation to generation. Particularly good examples of populations with such variation are insects such as the checkerspot butterfly (Murphy et al. 1990; see Figure 3.11), annual plants, and amphibians. In populations that show such extreme fluctuations, the effective population size is somewhere between the lowest and the highest number of individuals. However, the effective population size tends to be determined by the years with the smallest numbers; a single year of drastically reduced population numbers will substantially lower the value of N_e.

Bottlenecks and founder effects. When a population is greatly reduced in size, rare alleles in the population will be lost if no individuals possessing those alleles survive and reproduce. This phenomenon is known as a **population bottleneck**. With fewer alleles present and a decline in heterozygosity, the average fitness of the individuals in the population may decline. A special category of bottleneck, known as the **founder effect**, occurs when a few individuals leave a large population to establish a new population. The new population often has less genetic variability than the larger, original population and has a lower probability of persisting (Bryant et al. 1998).

The lions (*Panthera leo*) of Ngorongoro Crater in Tanzania provide a well-studied example of a genetic bottleneck (Packer 1992; 1997). The lion population consisted of 60–75 individuals until an outbreak of biting flies in 1962 reduced the population to 9 females and 1 male. Two years later, 7 additional males immigrated into the crater; there has been no further immigration since that time. The small number of founders, the isolation of the population, and variation in reproductive success among individuals apparently created a genetic bottleneck; even though the population increased to 125 animals by 1983, the population has since declined to 40 individuals (Packer, pers. comm. 2000). In comparison with the large Serengeti lion population nearby, the Crater lions show reduced genetic variability, high levels of sperm abnormalities (Figure 3.4), reduced reproductive rates, and increased cub mortality. Further genetic research on this important population has been stopped by the unwillingness of government officials to grant researchers permits to collect additional samples for analysis.

Population bottlenecks need not always lead to reduced heterozygosity. If a population expands rapidly in size after a temporary bot-

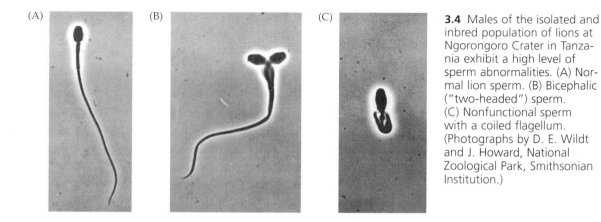

3.4 Males of the isolated and inbred population of lions at Ngorongoro Crater in Tanzania exhibit a high level of sperm abnormalities. (A) Normal lion sperm. (B) Bicephalic ("two-headed") sperm. (C) Nonfunctional sperm with a coiled flagellum. (Photographs by D. E. Wildt and J. Howard, National Zoological Park, Smithsonian Institution.)

tleneck, the former level of heterozygosity may be restored even though the number of alleles present is severely reduced (Nei et al. 1975; Allendorf and Leary 1986). An example of this phenomenon is the high level of heterozygosity found in the greater one-horned rhinoceros in Nepal. In the mid-1960s, this population consisted of fewer than 30 breeding individuals, but by 1988 it had recovered to almost 400 individuals (Dinerstein and McCracken 1990).

These examples demonstrate that effective population size is often substantially less than the total number of individuals in a population. Particularly where there is a combination of factors, such as fluctuating population size, numerous nonreproductive individuals, and an unequal sex ratio, the effective population size may be far lower than the number of individuals alive in a good year. For example, using special genetic techniques, the effective population size of a population of Chinook salmon was measured as 85, despite an apparent breeding population of 2000 adults. This disparity was attributed to the unequal breeding success of the adults (Bartley et al. 1992). A review of a wide range of wildlife studies revealed that the effective population size averaged only 11% of total population size. A population of 300 animals, seemingly large enough to maintain the population, might only have an effective population size of 33, indicating that it is in serious danger of loss of genetic variation and extinction (Frankham 1996). Such results demonstrate that simply maintaining large populations may not prevent the loss of genetic variation unless the effective population size is also large.

Demographic variation

In an idealized stable environment, a population will increase until it reaches the carrying capacity of the environment. At that point the average birth rate per individual equals the average death rate, and there is no net change in population size. In a real population, however, individuals do not usually produce the average number of offspring, but rather, may have no offspring, fewer than the average, or more than the average. Similarly, the average death rate in a population can be determined by studying large numbers of individuals. As long as population size is large, the average provides an accurate description of what is occurring in the population.

Once population size drops below about 50 individuals, individual variation in birth and death rates begins to cause the population size to fluctuate randomly up or down (Gilpin and Soulé 1986; Menges 1992). If population size fluctuates downward in any one year due to a higher than average number of deaths and a lower than average number of births, the resulting smaller population will be even more susceptible to demographic fluctuations in subsequent years. Random fluctuations upward in population size are eventually bounded by the carrying capacity of the environment, and the population may again fluctuate downward. Consequently, once a population becomes small due to habitat destruction and fragmentation, this demographic variation, also known as **demographic stochasticity**, becomes an important factor, and the population has a higher probability of going extinct due to chance alone (Lacy and Lindenmayer 1995). The chance of extinction is also greater in species that have low birth rates, such as elephants, because these species take longer to recover from a chance reduction in population size.

When populations drop below a critical number, there is also the possibility of a decline in birth rate due to an unequal sex ratio. For example, the last five surviving individuals of the extinct dusky sparrow (*Ammodramus maritimus nigrescens)* were all males, so there was no opportunity to establish a captive breeding program. Likewise, the last three individuals left in Illinois of the rare lakeside daisy (*Hymenoxys acaulis* var. *glabra)* are unable to produce viable seeds when cross-pollinating among themselves because they belong to the same self-infertile mating type (DeMauro 1993). In many animal species, small populations may be unstable due to the collapse of the social structure once the population falls below a certain number. Herds of grazing mammals and flocks of birds may be unable to find food and defend themselves against attack when their numbers fall below a certain level. Animals that hunt in packs, such as wild dogs and lions, may need a certain number of individuals to hunt effectively. Many animal species that live in widely dispersed populations, as do bears or whales, may be unable to find mates once the population density drops below a certain point. This is known as the **Allee effect**. In plant species, as population size decreases, the distance between plants increases; pollinating animals may not visit more than one of the isolated, scattered plants, resulting in a loss of seed production (Bawa 1990). This combination of random fluctuations in demographic characteristics, unequal sex ratios, decreased population density, and disruption of social behavior contributes to instabilities in population size, often leading to local extinction.

Environmental variation and catastrophes

Random variation in the biological and physical environment, known as **environmental stochasticity**, can also cause variation in the population size of a species. For example, the population of an endangered rabbit species might be affected by fluctuations in the population of a deer species that eats the same types of plants as the rabbits, by fluctuations in the population of a fox species that preys on the rabbits, and by the presence of parasites and diseases that affect the rabbits. Fluctuations in the physical environment might also strongly influence the rabbit population: rainfall during an average year might encourage plant growth and allow the population to increase, while dry years might limit plant growth and cause rabbits to starve.

Natural catastrophes at unpredictable intervals, such as droughts, storms, floods, earthquakes, volcanic eruptions, fires, and cyclical die-offs in the surrounding biological community, can also cause dramatic fluctuations in population levels. The likelihood of such extreme weather events will probably increase in coming decades as a consequence of global climate change. Natural catastrophes can kill part of a population or even eliminate the entire population from an area. Numerous examples exist of die-offs in populations of large mammals, including many cases in which 70%–90% of the population dies (Young 1994). Even though the probability of a natural catastrophe in any one year is low, over the course of decades and centuries, natural catastrophes have a high likelihood of occurring.

Modeling efforts by Menges (1992) and others have shown that random environmental variation is generally more important than random demographic variation in increasing the probability of extinction in populations of small to moderate size. In these models, environmental variation substantially increases the risk of extinction even in populations showing positive population growth under the assumption of a stable environment (Mangel and Tier 1994). In general, introducing environmental variation into population models—in effect making them more realistic—results in populations with lower growth rates, lower population sizes, and higher probabilities of extinction. In population models of a tropical palm species, the minimum viable population size—in this case, the number of individuals needed to give the population a 95% probability of persisting for 100 years—is about 48 mature individuals, taking into account demographic variation alone (Menges 1992). When low environmental variation is included, the minimum viable population size increases to 140 individuals; it increases to 380 individuals when moderate environmental variation is included. This result demonstrates that large populations must be protected to ensure species survival.

Extinction vortices

The smaller a population becomes, the more vulnerable it is to further demographic variation, environmental variation, and genetic factors, which all tend to reduce its size even more. This tendency of small populations to decline toward extinction has been likened to an **extinction vortex** (Gilpin and Soulé 1986). For example, a natural catastrophe, a new disease, or human disturbance could reduce a large population to a small size. This small population could then suffer from inbreeding depression, resulting in a lowered juvenile survival rate. This increased death rate could result in an even lower population size and even more inbreeding. Similarly, random demographic variation often reduces population size, resulting in even greater demographic fluctuations and a greater probability of extinction. These three factors—environmental variation, demographic variation, and loss of genetic variability—act together so that a decline in population size caused by one factor will increase the vulnerability of the population to the other factors (Figure 3.5). Once a population has declined, it will often become extinct unless

3.5 Extinction vortices progressively lower population size, leading to local extinctions of species. Once a population drops below a certain size, it enters a vortex, in which the factors that affect small populations tend to drive its size progressively lower. (After Gilpin and Soulé 1986; Guerrant 1992.)

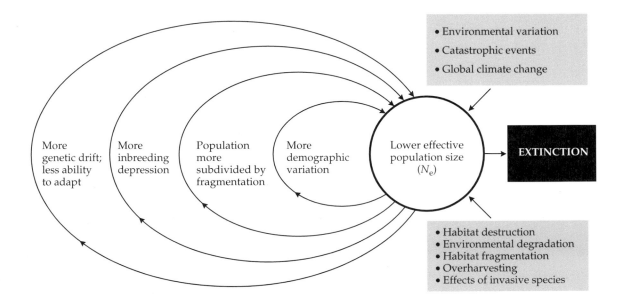

highly favorable conditions allow it to increase to a larger size. Such populations require a careful program of population and habitat management, as described later in the book, to reduce demographic and environmental variation and thus minimize the effects of small population size.

Natural History and Ecology

The key to protecting and managing a rare or endangered species is an understanding of its biological relationship to its environment and the status of its populations (Schaller 1993). This information is generally called the **natural history**, or sometimes simply the **ecology**, of the species. With information concerning a rare species' natural history, managers are able to make more effective efforts to maintain the species and identify the factors that place it at risk of extinction (Gilpin and Soulé 1986).

The following are categories of ecological questions that need to be answered in order to design and implement effective population-level conservation efforts. For most species, only a few of these questions can be answered without further investigation. Decisions on management, however, often have to be made before this information is available or while it is being gathered. The exact types of information gathered will obviously depend on the characteristics of the species.

- *Environment.* What are the habitat types in which the species is found, and how much area is there of each? How variable is the environment in time and space? How frequently is the environment affected by catastrophic disturbance? How have human activities affected the environment?

- *Distribution.* Where is the species found in its habitat? Does the species move and migrate among habitats, or to different geographical areas, over the course of a day or over a year? How efficient is the species at colonizing new habitats? How have human activities affected the distribution of the species?

- *Biotic interactions.* What types of food and other resources does the species need? What other species compete with it for those resources? What predators, pests, and parasites affect its population size?

- *Morphology.* How do the shape, size, color, and surface texture of individuals allow the species to exist in its environment?

- *Physiology.* What quantities of food, water, minerals, and other necessities does an individual need to survive, grow, and reproduce? How efficient is an individual at using its resources? How vulnerable is the species to extremes of climate, such as heat, cold, wind, and rain?

- *Demography.* What is the current population size, and what was it in the past? Are the numbers of individuals stable, increasing, or decreasing?

- *Behavior.* How do the actions of an individual allow it to survive in its environment? How do individuals in a population mate and produce offspring? In what ways do individuals of the species interact among themselves, either cooperatively or competitively?

- *Genetics.* How much of the variation among individuals in morphological and physiological characteristics is genetically controlled?

Gathering ecological information

The basic information needed for an effort to conserve a species or determine its status can be obtained from three major sources.

- *Published literature.* Library indexes such as *Biological Abstracts* or *Zoological Record*, often accessible by computer, provide easy access to a variety of books, articles, and reports. This literature may contain records of previous population sizes and distributions that can be compared with the current status of the species (Greenberg and Droege 1999). The Internet provides ever-increasing access to databases, electronic bulletin boards, and specialized discussion groups and subscription databases such as the *Web of Science*. Sometimes sections of the library will have related material shelved together, so that finding one book leads to other books. Also, once one key reference is obtained, its bibliography can often be used to discover earlier useful references. The *Science Citation Index*, available in many libraries and on the *Web of Science*, is another valuable tool for tracing the literature forward in time; for example, by looking in the current *Science Citation Index* for the name of W. K. Kenyon, who wrote several important scientific papers on the Hawaiian monk seal between 1959 and 1981, recent papers on the Hawaiian monk seal that cited his works can be located.

- *Unpublished literature.* A considerable amount of information in conservation biology is contained in unpublished reports by individuals, government agencies, and conservation organizations. This so-called "gray literature" is sometimes cited in published literature or mentioned by leading authorities in conversations and lectures; it is increasingly being posted on the Internet. Often, such reports can be obtained through direct contact with the author or with conservation organizations. (A list of such organizations is given in the Appendix.)

- *Fieldwork.* The natural history of a species usually must be learned through careful observations in the field. Fieldwork is usually necessary because only a tiny percentage of the world's species have been studied, and because the ecology of many species changes from one place to another. Only in the field can the conservation status of a species, and its relationships to the biological and physical environment, be determined. While much information can be obtained through careful observation, many of the other techniques used in the field are technical and are best learned by study under the supervision of an expert or by reading manuals (for example, Rabinowitz 1993; Kricher 1998; Wilson et al. 1996).

Monitoring populations

The way to learn the status of a rare species of special concern is to census the species in the field and monitor its populations over time. By repeatedly censusing a population on a regular basis, changes in the population over time can be determined (Schemske et al. 1994; Primack 1998b). Long-term census records can help to distinguish bona fide population trends of increase or decrease—possibly caused by human disturbance—from short-term fluctuations due to variations in weather or unpredictable natural events (Pechmann et al. 1991; Cohn 1994). Monitoring is effective at showing the response of a population to a change in its environment; for example, through monitoring, a decline in an orchid species was shown to be connected with heavy cattle grazing in its habitat (see below). Monitoring efforts can also be targeted at particularly sensitive species, such as butterflies, using them as **indicator species** of the long-term stability of ecological communities (Sparrow et al. 1994).

Monitoring studies have been increasing dramatically over the last decade as government agencies and conservation organizations become more concerned with protecting rare and endangered species (Goldsmith 1991). A review of projects monitoring rare and endangered plants in the United States showed a phenomenal increase in the number initiated from 1974 to 1984: only one project was launched in the three years from 1974 to 1976, while more than 120 were started from 1982 to 1984 (Palmer 1987). The most common types of monitoring projects were inventories (40%) and population demographic studies (40%); survey studies were somewhat less frequently used (20%).

An **inventory** is simply a count of the number of individuals present in a population. By repeating an inventory over successive time intervals, it can be determined whether the population is stable, increasing, or decreasing in number. The inventory is an inexpensive and straightforward method. It might answer such questions as: How many individuals exist in the population at present? Has the population been stable in number during the period for which inventory records exist? Inventories conducted over a wider area can help to determine the range of a species and its areas of local abundance.

A **population survey** involves the use of a repeatable sampling method to estimate the density of a species in a community. An area can be divided into sampling segments, and the number of individuals in each segment counted. These counts can then be averaged and used to estimate the actual population size. Survey methods are used when a population is very large or its range extensive. Survey methods are particularly valuable when stages in a species' life cycle are inconspicuous, tiny, or hidden, such as the seed and seedling stages of many plants or the larval stages of aquatic invertebrates.

Demographic studies follow known individuals in a population to determine their rates of growth, reproduction, and survival. Individuals of all ages and sizes must be included in such a study. Either the whole population or a subsample can be followed. In a complete population study, all individuals are counted, aged if possible, measured for size, sexed, and tagged or marked for future identification; their positions on the site are mapped, and tissue samples are sometimes collected for genetic analysis. The techniques used to conduct a population study vary depending on the characteristics of the species and the purpose of the study. Each discipline has its own technique for following individuals over time: ornithologists band birds' legs, mammalogists often attach tags to animals' ears, and botanists attach aluminum tags to trees (Figure 3.6; see Goldsmith 1991). Information from demographic studies can be used in life history formulas to calculate the rate of population change and to identify vulnerable stages in the life cycle (Caswell 1989; Tuljapurkar and Caswell 1997).

Demographic studies can provide information on the age structure of a population. A stable population typically has an age distribution with a characteristic ratio of juveniles, young adults, and older adults. An absence or low number of any age class, particularly of juveniles, may indicate that the population is in danger of declining. Similarly, a large number of juveniles and young adults may indicate that the population is stable or even expanding. Careful analysis of long-term data, or of changes in the population over time, is often needed to distinguish short-term fluctuations from long-term trends.

Demographic studies can also reveal the spatial characteristics of a species, which can be very important in maintaining the viability of separate populations. The number of populations of the species, movement among the populations, and the stability of the populations in space and time are all important considerations, particularly for species that occur in an aggregate of temporary or fluctu-

3.6 Monitoring populations requires specialized techniques suited to each species. (A) An ornithologist checks the health and weight of a piping plover on Cape Cod. Note the identification band on the bird's leg. (Photograph by Laurie McIvor.) (B) Botanists monitor tagged lady's slipper orchid plants (*Cypripedium acaule*) for their changes in leaf size and number of flowers over a ten-year period. As shown here, individual leaves are monitored for their rates of carbon dioxide uptake, a measure of photosynthetic rate and an index of plant health. Note the numbered aluminum tag, anchored to the ground by a wire; these tags have remained in place for 16 years. (Photograph by Richard Primack.) (C) Censusing fish populations on a tropical reef ecosystem. (Photograph © Simon Jennings.)

ating populations linked by migration, known as a **metapopulation** (see below). Some examples of monitoring studies follow.

- *Hawaiian monk seals.* Population inventories of the Hawaiian monk seal (*Monachus schauinslandi*) documented a decline from almost 100 adults in the 1950s to fewer than 14 in the late 1960s (Figure 3.7; Gerrodette and Gilmartin 1990). The number of pups similarly declined during this period. On the basis of these trends, the Hawaiian monk seal was declared an endangered species under the U.S. Endangered Species Act in 1976, and con-

3.7 An inventory of the Hawaiian monk seal population on Green Island, Kure Atoll (black trace), revealed that the population declined sharply following the opening of a Coast Guard station. The population inventory was plotted from either a single count, the mean of several counts, or the maximum of several counts. (After Gerrodette and Gilmartin 1990.)

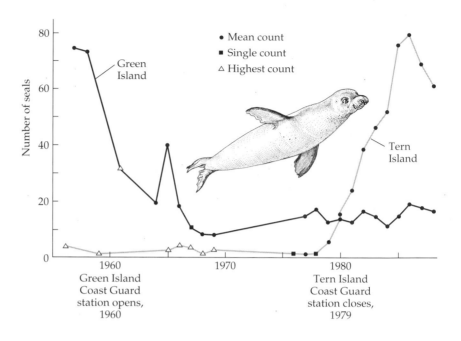

servation efforts were implemented at Tern Island that reversed the trend (Ackerman 1992).

- *Marine mollusks.* In Transkeii, South Africa, people living on the coast collect and eat marine mollusks, such as the brown mussel, the abalone, and the turban shell (Lasiak 1991). A survey method was used to determine whether traditional collecting methods are likely to deplete shellfish populations: The frequency and size distributions of mollusks were compared in protected and exploited rocky areas. The survey determined that although collection depleted the adult populations in exploited areas, they were quickly replaced by larvae, probably due to immigration from nearby protected areas and adjacent, inaccessible subtidal areas.

- *Early spider orchid.* The early spider orchid (*Ophrys sphegodes*) has shown a substantial decline in range during the last half of the twentieth century in Britain. A nine-year demographic study of this species showed that the plants were unusually short-lived for perennial orchids, with only half of the individuals surviving beyond two years (Hutchings 1987). This short half-life makes the species unusually vulnerable to unfavorable conditions. In one population in which the species was declining in numbers, demographic analysis highlighted soil damage caused by cattle grazing as the key element in the population decline. A change that allows only sheep grazing during times when the plants are not flowering and fruiting has enabled the population to make a substantial recovery.

Monitoring studies are playing an increasingly important role in conservation biology. Monitoring has a long history in temperate countries, particularly Britain (Goldsmith 1991). In North America, the Breeding Bird Survey has been censusing bird abundance at about 1000 locations for the past 30 years, and this information is now being used to determine the stability of migrant songbird populations over time (James et al. 1996). Some of the most elaborate monitoring projects have established permanent research plots in tropical forests, such as the 50 ha site at Barro Colorado Island in Panama, to monitor changes in species and communities over time (Condit et al. 1992). These studies have shown that many tropical tree and bird species are more dynamic in numbers than had previously been suspected (Laurance and Bierregaard 1997), suggesting that MVP estimates may need to be revised upward.

Population viability analysis

Population viability analysis (**PVA**) is an extension of demographic analysis that focuses on determining whether a species has the ability to persist in an environment (Soulé 1990; Ruggiero et al. 1994; Akçakaya et al. 1999). PVA examines the range of requirements that a species has and the resources available in its environment in order to identify vulnerable stages in its natural history. PVA can be useful in understanding the effects of habitat loss, habitat fragmentation, and habitat degradation on a rare species. Although PVA is still being developed as an approach for predicting species persistence and does not yet have a standard methodology or statistical framework (Burgman et al. 1993; Lacy and Lindemayer 1995), its methods of systematically and comprehensively examining species data are logical extensions of natural history research and demographic studies. Such attempts to use statistics to predict future trends in population sizes must be used with caution, along with a large dose of common sense, because conditions may change and conservation efforts may prove successful or ineffective (Mann and Plummer 1999).

Attempts to utilize population viability analysis have already begun. One of the most thorough examples of PVA, combining genetic and demographic analyses, is a study of the Tana River crested mangabey (*Cercocebus galeritus galeritus*), an endangered primate confined to the floodplain forests in a nature reserve along the Tana River in eastern Kenya (Figure 3.8; Kinnaird and O'Brien 1991). As its habitat has been reduced in area and fragmented by agricultural activities in the last 15 to 20 years, the species has experienced

3.8 The Tana River National Primate Reserve, Kenya; within the reserve, remaining forested patches located along the river are shown as shaded areas. The Tana River crested mangabey occurs only in this region of East Africa, and is increasingly endangered by forest fragmentation and human encroachment on its habitat. (After Kinnaird and O'Brien 1991.)

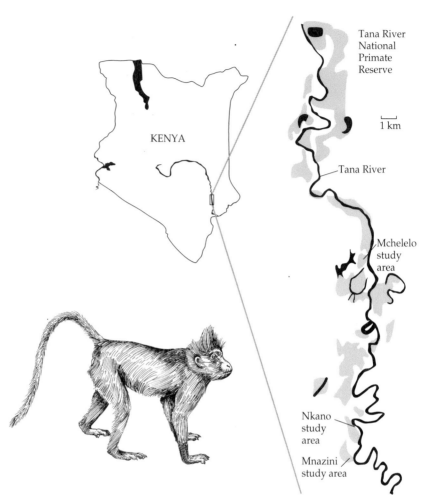

a decline in overall population size of about 50%, as well as a decline in the number of groups. While the number of mangabeys in 1989 was about 700, the effective population size was only about 100 due to a large proportion of nonreproductive individuals and variation in the number of offspring produced by different individuals. With such a low effective population size, the mangabey is in danger of losing significant amounts of its genetic variability. To maintain an effective population size of 500 individuals, the number considered sufficient to preserve genetic variability, a population of about 5000 mangabeys would have to be sustained. In addition, a demographic analysis of the population suggests that in the current situation, the probability of the population going extinct over the next 100 years is 40%. To ensure that the population has a 95% probability of persisting for 100 years, based on demographic factors alone, the population size would have to be almost 8000 individuals.

Both the genetic and the demographic analyses suggest that the long-term future of the present mangabey population is bleak. Given the restricted range and habitat of the species and the growing human population in the area, a goal of increasing the population size to 5000–8000 individuals is probably unrealistic. A management plan that combines increases in the area of protected forests, enrichment plantings of existing forests to increase the number of mangabey food plants, and establishment of corridors to facilitate movement between forest fragments might increase the survival probability of the Tana River crested mangabey.

PVA has also played a role in the conservation efforts on behalf of the African elephant, which have taken on international importance because of the species' precipitous decline in numbers and its symbolic importance as a representative of wildlife throughout the world. A population viability analysis of elephant populations on semiarid land at Tsavo National Park in Kenya indicated that a minimum reserve size of about 2500 km^2 is needed to attain a 99% probability of population persistence for 1000 years (Figure 3.9; Armbruster and Lande 1993). At densities of about 12 animals per

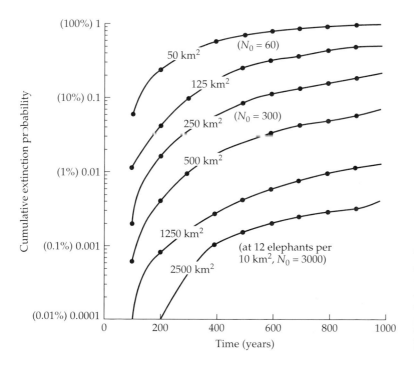

3.9 Cumulative probability of extinction (log scale) over time for elephant populations in protected areas of different sizes. With a density of 12 elephants per 10 km^2, a 2500 km^2 protected area has an initial population (N_0) of 3000 elephants; the probability of extinction in 100 years is close to 0%, and in 1000 years is just 0.4%. A population in a protected area of 250 km^2 with an initial population of 300 elephants has a 20% probability of extinction in 1000 years. (After Armbruster and Lande 1993.)

10 km², this translates into an initial population size of about 3000 animals. At this reserve size, the population could tolerate a modest degree of harvesting without substantially increasing its probability of extinction.

The metapopulation

Over the course of time, populations of a species may become extinct on a local scale, and new populations may form on other nearby, suitable sites. Many species that live in ephemeral habitats, such as herbs found on frequently flooded stream banks or in recently burned forests, are best characterized by **metapopulations** ("a population of populations") made up of a shifting mosaic of temporary populations linked by some degree of migration. In these species, every population is short-lived, and the distribution of the species changes dramatically with each generation. In other species, the metapopulation may be characterized by one or more **core** populations, with fairly stable numbers, and several **satellite areas** with fluctuating populations. Populations in the satellite areas may become extinct in unfavorable years, but the areas are recolonized ("rescued") by migrants from the core population when conditions become more favorable (Figure 3.10) (Hanski et al 1996; Hanski and Simberloff 1997). The target of a population study is typically one or several populations, but an entire metapopulation

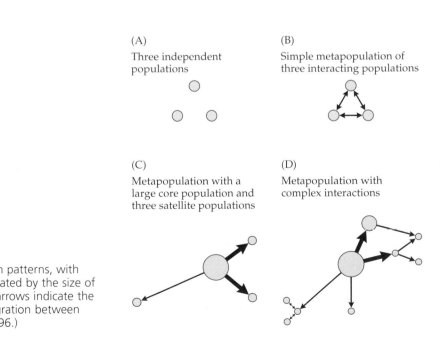

(A)
Three independent populations

(B)
Simple metapopulation of three interacting populations

(C)
Metapopulation with a large core population and three satellite populations

(D)
Metapopulation with complex interactions

3.10 Possible metapopulation patterns, with the size of a population indicated by the size of its representative circle. The arrows indicate the direction and intensity of migration between populations. (After White 1996.)

may need to be studied if it would result in a more accurate portrayal of the species.

Metapopulation models have the advantage of recognizing that local populations are dynamic and that there is movement of organisms from one local population to another. The recognition that infrequent colonization events and migrations are occurring also allows biologists to consider the impact of founder effects and genetic drift on the species. The following are two examples in which the metapopulation approach has proved to be more useful than a single-population description in understanding and managing species.

The endemic Furbish's lousewort (*Pedicularis furbishiae*) occurs along a river in Maine that is subject to periodic flooding (Menges 1990). Flooding often destroys some existing populations, but it also creates exposed riverbank habitat suitable for the establishment of new populations. Studies of any single population would give an incomplete picture of the species because the current populations are short-lived. The metapopulation is really the appropriate unit of study, and the watershed is the appropriate unit of management.

The checkerspot butterfly (*Euphydryas* spp.) has been extensively studied in California (Figure 3.11; Murphy et al. 1990). Individual butterfly populations often become extinct, but dispersal and colonization of unoccupied habitat allow the species to survive. Environmental stochasticity and lack of habitat variation at a particular site often cause extinctions of local populations. The largest and most persistent populations are found in large areas that have both moist, north-facing slopes and warmer, south-facing slopes. Butterflies migrating out from these core populations often colonize the unoccupied satellite areas.

3.11 Studies of the bay checkerspot butterfly (*Euphydryas editha bayensis*) have been used to demonstrate the metapopulation approach. Core populations of this species colonize unoccupied satellite areas during years when conditions are favorable. (Photograph courtesy of Dennis Murphy, Stanford University.)

In metapopulations, destruction of the habitat of one central, core population could result in the extinction of numerous satellite populations over a wider area which depend on the core population for periodic colonization. Also, human disturbances that inhibit migration, such as fences, roads, and dams, can reduce the rate of migration among habitat patches and so reduce or even eliminate the probability of recolonization after local extinction (Lindenmayer and Lacy 1995). Habitat fragmentation resulting from these and other activities sometimes has the effect of transforming a large continuous population into a metapopulation in which small residual populations occupy habitat fragments. When population size in these fragments is small and the rate of migration among fragments is low, populations within each fragment will gradually go extinct and recolonization will not occur. Effective management of a species often requires an understanding of these metapopulation dynamics.

Long-term monitoring of species and ecosystems. Long-term monitoring of ecosystem processes (temperature, rainfall, humidity, soil acidity, water quality, discharge rates of streams, soil erosion, etc.), communities (species present, amount of vegetative cover, amount of biomass present at each trophic level, etc.), and population numbers (number of individuals present of particular species) is necessary because it is otherwise difficult to distinguish normal year-to-year fluctuations from long-term trends (Magnuson 1990; Primack 1992). For example, many amphibian, insect, and annual plant populations are highly variable from year to year, so many years of data are required to determine whether a particular species is actually declining in abundance or merely experiencing a number of low-population years in accord with its regular pattern of variation. For example, 40 years of observation of populations of two flamingo species in South Africa revealed that large numbers of chicks fledged only in years with high rainfall (Figure 3.12). The number of chicks fledging currently is much lower than in the past, indicating that the species may be heading toward local extinction. Monitoring is particularly important in integrated conservation and development projects, in which the long-term protection of biological diversity is an important goal (Kremen et al. 1994; Bawa and Menon 1997). Monitoring allows managers to determine if the goals of their projects are being achieved, or if adjustments have to be made in management plans.

The fact that environmental effects may lag for many years behind their initial causes creates a challenge to understanding change in ecosystems. For example, acid rain and other components of air pollution may weaken and kill trees over a period of decades, increasing the amount of soil erosion into adjacent streams and ultimately mak-

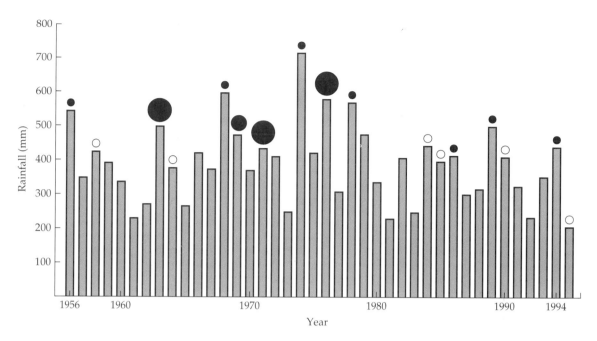

3.12 The bars show rainfall data from Etosha National Park for the years 1956 to 1995. The flamingo breeding events that occurred in those years are indicated by circles. Open circles indicate failed breeding events: eggs were laid but no chicks hatched. The small, medium, and large, black circles indicate, respectively, that fewer than 100 chicks hatched, hundreds of chicks hatched, and thousands of chicks hatched. The last large hatching occurred in 1976. (From Simmons 1996.)

ing the aquatic environment unsuitable for the larvae of certain rare insect species. In such a case, the cause (air pollution) may have occurred decades before the effect (insect decline) is detectable. Acid rain, global climate change, vegetation succession, nitrogen deposition, and invasions of exotic species are all examples of processes that cause long-term changes in biological communities, but are often hidden from our short-term perspective. Some long-term information is available from weather stations, annual census counts of birds, forestry plots, water authorities, and old photographs of vegetation, but the number of long-term monitoring efforts for biological communities is inadequate for most conservation purposes. To remedy this situation, many scientific research stations have begun to implement programs for monitoring ecological change over the course of decades and centuries. One such program is the system of 172 Long-Term Ecological Research (LTER) sites established by the U.S. National Science Foundation (Swanson and Sparks 1990). Another proposed U.S. program is the Biodiversity Observatory Network of 20 sites (Dalton 1999). These sites will provide an early warning system for disruption or decline of ecosystem functions.

Establishment of New Populations

Instead of only passively observing endangered species as they decline toward extinction, many conservation biologists have begun to develop approaches to save these species. Some exciting new methods are being developed to establish new wild and semi-wild populations of rare and endangered species and to increase the size of existing populations (Gipps 1991; Bowles and Whelan 1994). These experiments offer the hope that species now living only in captivity can regain their ecological and evolutionary roles within the biological community. Populations in the wild may have less chance of being destroyed by catastrophes (such as epidemics or wars) than confined captive populations. Further, simply increasing the number and size of populations of a species generally lowers its probability of extinction.

Such establishment programs are unlikely to work effectively, however, unless the factors leading to the decline of the original wild populations are clearly understood and then eliminated, or at least controlled. For example, if an endemic bird species has been hunted nearly to extinction in the wild by local villagers, its nesting areas damaged by development, and its eggs eaten by an exotic species, these issues have to be addressed as an integral part of a reestablishment program. Simply releasing captive-bred birds into the wild without discussions with local people, a change in land use patterns, and control of exotic species would result in a recurrence of the original situation.

Three basic approaches have been used to establish new populations of animals and plants. A **reintroduction program** involves releasing captive-bred or wild-collected individuals into an area of their historic range where the species no longer occurs. (There is confusion among the terms denoting the reintroduction of populations, and sometimes these programs are also called "reestablishments," "restorations," or "translocations.") The principal objective of a reintroduction program is to create a new population in the original environment. For example, a program initiated in 1995 to reintroduce gray wolves into Yellowstone National Park aims to restore the equilibrium of predators and herbivores that existed prior to human intervention in the region (Figure 3.13). Individuals frequently are released at or near the site where they or their ancestors were collected to ensure genetic adaptation to the site. Individuals are also sometimes released elsewhere within the range of the species when a new protected area has been established, when an existing population is under a new threat and will no longer be able to survive in its present location, or when natural or artificial barriers to the normal dispersal tendencies of the species exist.

Two other distinct types of release programs are also being used. An **augmentation program** involves releasing individuals into an existing population to increase its size and gene pool. These released individuals may be wild individuals collected elsewhere or individ-

3.13 A gray wolf stalks an elk herd in Yellowstone National Park. As a result of the wolves' reintroduction to the park, elk have changed their behavior, congregating in dense herds and becoming more alert to danger. As a keystone predator, the activity of wolves has already altered the behavior and population numbers of many other species, including grizzly bears, coyotes, and carrion beetles. (Photograph by Bill Campbell.)

uals raised in captivity. One special example of augmentation is a "headstarting" approach in which marine turtle hatchlings are raised in captivity during their vulnerable young stage, and then are released into the wild. An **introduction program** involves moving animals and plants to areas outside their historic range in the hope of establishing new populations. Such an approach may be appropriate when the environment within the historic range of a species has deteriorated to the point at which the species can no longer survive there, or when the factor causing the original decline is still present, making reintroduction impossible. A planned introduction of a species to new sites needs to be carefully researched to ensure that the species does not damage its new ecosystem or harm populations of any local endangered species. Care must also be taken to ensure that released individuals have not acquired diseases while in captivity that could spread to and decimate wild populations.

Considerations for successful programs

Programs to establish new populations are often expensive and difficult; they require a serious, long-term commitment. The programs implemented to capture, raise, monitor, and release California condors, peregrine falcons, and black-footed ferrets, for instance, have cost tens of millions of dollars and required years of work. When the animals are long-lived, the program itself may have to last for many years before its outcome is known. Decisions on initiating reintroduction programs can also become highly emotional public issues, as evidenced by the California condor, black-footed ferret, grizzly bear, and gray wolf programs in the United States, and comparable programs in European countries. Programs can be attacked as a waste of money ("Millions of dollars for a few ugly birds!"), unnecessary ("Why do we need wolves here when there are so many elsewhere?"), poorly run ("Look at all of the ferrets that died of disease in captivity!"), or unethical ("Why can't the last animals just be allowed to live out their lives in peace without being captured and put into zoos?"). The answer to all of these criticisms is straightforward: Well-run, well-designed captive breeding and reintroduction programs are the best hope for preserving a species that is about to become extinct in the wild or is in severe decline. A crucial element in many reintroduction programs will be public relations efforts to explain the need for and the goals of the program to local people and convince them to support it, and perhaps even take pride in it, or at least not oppose it (Reading and Kellert 1993; Milton et al. 1999). Providing incentives to the local community as part of the program often is more successful than rigid enforcement of restrictions and laws.

Released animals may require special care and assistance during and immediately after release; this approach is known as "soft release." Animals may have to be fed and sheltered at the release point until they are able to subsist on their own, or they may need to be caged temporarily at the release point and released gradually, so that they can become familiar with the area (Figure 3.14). Social groups abruptly released from captivity (a "hard release") may explosively disperse in different directions and away from the pro-

3.14 Cages allow black-footed ferrets (*Mustela nigrepes*) to become familiar with the range where they will eventually be released. The ferrets' caretaker is wearing a mask to reduce the chance of the ferrets being exposed to human disease. (Photograph by LuRay Parker, Wyoming Game and Fish Department.)

tected area, resulting in a failed establishment effort. Intervention may be necessary if the animals appear to be unable to survive, particularly during episodes of drought or low food abundance. In such cases a decision has to be made whether it is better to give the animals occasional temporary help to get them established or to force them to survive on their own. Reintroduction efforts often involve a long-term commitment to conservation, extending over decades.

Successful reintroduction programs often have considerable educational value. In Brazil, efforts to preserve the golden lion tamarin (*Leontopithecus rosalia*) through conservation and reintroduction have become a rallying point in attempts to protect the last remaining fragments of the Atlantic Coast forest (Dietz et al. 1994). In Oman, captive-bred Arabian oryx (*Oryx leucoryx*) were successfully reintroduced into desert areas, creating an important national symbol and a source of employment for the local Bedouins who run the program (Stanley-Price 1989), though this population is under increasing threat from poaching. Attempts are also under way to establish new populations of endangered insects, which is appropriate considering that there are so many species of insects, only some of which are well known to the public (Samways 1994).

Establishment programs for common game species have always been widespread and have contributed a great deal of knowledge to the new programs being developed for threatened and endangered species. A detailed study that examined 198 bird and mammal establishment programs conducted between 1973 and 1986 found a number of significant generalizations (Griffith et al. 1989). The reported success of programs in establishing new populations is:

- Greater for game species (86%) than for threatened, endangered, and sensitive species (44%)

- Greater for release in excellent quality habitat (84%) than in poor quality habitat (38%)

- Greater in the core of the historic range (78%) than at the periphery of and outside the historic range (48%)

- Greater with wild-caught (75%) than with captive-bred animals (38%)

- Greater for herbivores (77%) than for carnivores (48%)

For these bird and mammal species, the probability of establishing a new population increased with the number of animals being released up to about 100 animals. Releasing more than 100 animals did not further enhance the probability of success.

A second survey of projects (Beck et al. 1994) used a more restricted definition of reintroduction: the release of captive-born birds and mammals within the historical range of the species. A program was judged a success if there was a self-maintaining population of 500 individuals. Using these precise definitions, only 16 out of 145 reintroduction projects were judged successful—a dramatically *lower* rate of success than the earlier survey. According to this new study, the key to success involves releasing large numbers of animals over many years. Reintroductions involving fish, reptiles, and amphibians also have a low rate of success, perhaps due to specialized habitat requirements (Dodd and Seigel 1991; Hendrickson and Brooks 1991; Minckley 1995). Clearly, monitoring and evaluating ongoing and future programs are crucial in order to determine whether efforts to establish new populations are achieving their stated goals.

Social behavior of released animals

Successful reintroduction, augmentation, and introduction programs need to consider the social organization and behavior of the animals that are being released (Caro 1998). When social animals (particularly mammals and some birds) grow up in the wild, they learn about their environment and how to interact socially from other members of their species. Animals raised in captivity may lack the skills needed to survive in their natural environment, as well as the social skills necessary to find food cooperatively, sense danger, find mating partners, and raise young. To overcome these socialization problems, captive-raised mammals and birds may require extensive training before as well as after their release into the environment (Kleiman 1989; Curio 1996; Clemmons and Buchholz 1997). Captive chimpanzees, for instance, have been taught how to use twigs to feed on termites and how to build nests. Red wolves are taught how to kill live prey. Captive animals are taught to fear potential predators by being frightened in some way when a dummy predator is shown.

Social interaction is one of the most difficult behaviors for people to teach to captive-bred mammals and birds because, for most species, the subtleties of social behavior are poorly understood. Nevertheless, some successful attempts have been made to socialize captive-bred mammals (Valutis and Marzluff 1999). In some cases, humans mimic the appearance and behavior of wild individuals. This method is particularly important when dealing with very young animals, which need to learn to identify with their own species rather than with a foster species or with humans. Captive-bred California condor hatchlings were originally unable to learn the behaviors of their wild relatives because they had imprinted on their human keepers. Newly hatched condors are now fed with condor hand puppets and kept from the sight of zoo visitors (Figure

3.15 California condor chicks (*Gymnogyps californianus*) raised in captivity are fed by researchers using hand puppets that look like adult birds. Conservation biologists hope that minimizing human contact with the birds will improve their chances of survival when they are returned to the wild. (Photograph by Mike Wallace, The Los Angeles Zoo.)

3.15). In other cases, wild individuals are used as "instructors" for captive individuals of the same species. Wild golden lion tamarins are caught and held with captive-bred tamarins to form social groups that are then released together, in the hope that the captive-bred tamarins will learn from the wild ones. When captive-bred animals are released into the wild, they sometimes join existing social groups or mate with wild animals and thereby gain some knowledge of their environment (Figure 3.16). The development of social relationships with wild animals may be crucial to the success of the captive-bred animals once they have been released.

Establishing new plant populations

Efforts at establishing new populations of rare and endangered plant species are fundamentally different from attempts using terrestrial vertebrate animal species. Animals can disperse to new locations and actively seek out the microsite conditions that are most suitable for them. In the case of plants, seeds are dispersed to new sites by such agents as wind, animals, and water (Primack and Miao 1992; Falk et al. 1996; Primack and Drayton 1997). Once a seed lands on the ground, it is unable to move farther, even if a suitable microsite exists just a few centimeters away. The immediate microsite is crucial for plant survival: If the environmental condi-

3.16 An experimental population of golden lion tamarins in Brazil initially consisted almost entirely of reintroduced, captive-born animals, but is now primarily wild-born animals, which may be due in some part to wild tamarins "teaching" captive-born tamarins to survive in the wild. This seems to indicate a successful program and a population that could soon become self-sustaining. (From Beck and Martins 1995.)

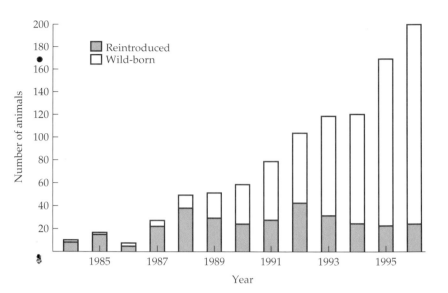

tions are too sunny, too shady, too wet, or too dry, either the seed will not germinate or the resulting seedling will die. Plant ecologists are currently investigating the effectiveness of site treatment, such as burning the site, removing competing vegetation, digging up the ground, and excluding grazing animals, as a means of enhancing population establishment (Figure 3.17). In one case, a reintroduction of the rare large-flowered fiddleneck (*Amsickia grandiflora*) in a California grassland involved burning and use of selective herbicide to

3.17 A variety of methods are being investigated to create new populations of rare wildflower species on U.S. Forest Service land in South Carolina. Seeds are being planted in a pine forest from which the oak understory has been removed. Wire cages will be placed over some plantings to determine if excluding rabbits, deer, and other animals will help in plant establishment. (Photograph by Richard Primack.)

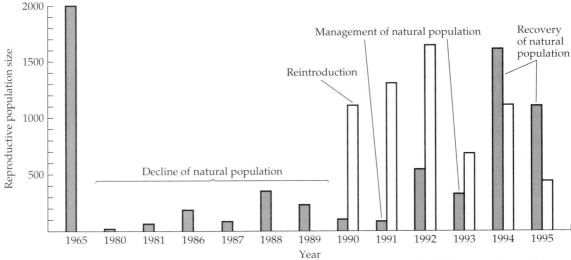

3.18 A natural population (dark gray bars) of *Amsickia grandiflora*, an endangered annual plant of northern California, was in severe decline during the 1980s as a result of competition with exotic annual grasses. A reintroduction was carried out at a second site starting in 1990 (white bars), and was combined with various management treatments to eliminate exotics. Successful treatments were then applied to the natural population in 1991 and 1993, with major increases in plant numbers seen in 1992 and 1994. (From Guerrant and Pavlik 1998.)

remove exotic grasses. These techniques were then successfully applied to a declining natural population, resulting in a dramatic increase in the number of individuals (Figure 3.18).

Populations of rare and endangered plant species typically fail to become established from introduced seeds at most sites that appear to be suitable for them (Primack 1996). To increase their chances of success, botanists often germinate seeds in controlled environments and grow the young plants in protected conditions. Only after the plants are past the fragile seedling stage are they transplanted into the field. In other cases, plants are dug up from an existing wild population (usually either one that is threatened with destruction or one in which removing a small percentage of the plants will not apparently harm the population), then transplanted into an unoccupied but apparently suitable site. While such transplantation methods have a good chance of ensuring that a species survives at its new location, they do not mimic a natural process, and new populations sometimes fail to produce the seeds and seedlings needed to form the next generation. As research on this rapidly developing topic is published and synthesized, the chances of success will hopefully improve.

New populations and the law

Reintroduction, introduction, and augmentation programs will increase in the coming years as the biological diversity crisis eliminates more species and populations from the wild. Many of the reintroduction programs for endangered species will be mandated by official recovery plans set up by national governments (Tear et al. 1995). However, establishment programs, as well as research in general on endangered species, are increasingly being affected by legislation that restricts the possession and use of endangered species (Reinartz 1995; Falk et al. 1996). If government officials rigidly apply these laws to scientific research programs, which was certainly not the original intent of the legislation, the creative insights and new approaches coming out of these programs could be stifled (Ralls and Brownell 1989). New scientific information is central to establishment programs and other conservation efforts. Government officials who block reasonable scientific projects may be doing a disservice to the organisms they are trying to protect. The harm to endangered species that could be caused by carefully planned scientific research is relatively insignificant in comparison with the actual massive loss of biological diversity being caused by habitat destruction and fragmentation, pollution, and overexploitation. Conservation biologists must be able to explain the benefits of their research programs in a way that government officials and the general public can understand, and must address the legitimate concerns of those groups (Farnsworth and Rosovsky 1993).

Experimental populations of rare and endangered species successfully created by introduction and reintroduction programs are given various degrees of legal protection (Falk and Olwell 1992). "Experimental essential" populations are regarded as critical to the survival of the species and are as rigidly protected as naturally occurring populations. "Experimental nonessential" populations are not protected under the law; designating populations as nonessential often helps to overcome the fear of local landowners that having an endangered species on their property will restrict how their land can be managed and developed. Legislators and scientists alike must understand that the establishment of new populations through reintroduction programs in no way reduces the need to protect the original populations of endangered species; the original populations are more likely to have the most complete gene pool and the most intact interactions with other members of the biological community. In many cases, proposals to create new habitats and populations elsewhere are made to mitigate or compensate for damage to species and habitats that has already occurred or is about to occur. Given the poor success of most attempts to create new populations of rare species, protection of existing populations should be given the highest priority.

Ex Situ Conservation Strategies

The best strategy for the long-term protection of biological diversity is the preservation of natural communities and populations in the wild, known as **in situ** or **on-site preservation**. Only in the wild are species able to continue the process of evolutionary adaptation to a changing environment within their natural communities. However, for many rare species, in situ preservation is not a viable option in the face of increasing human disturbance. If a remnant population is too small to persist, or if all the remaining individuals are found outside of protected areas, then in situ preservation may not be effective. In such circumstances it is likely that the only way to prevent the species from becoming extinct is to maintain individuals in artificial conditions under human supervision (Kleiman et al. 1996). This strategy is known as **ex situ** or **off-site preservation**. Already a number of animal species are extinct in the wild but survive in captive colonies, such as the Père David's deer (*Elaphurus davidianus*) (Figure 3.19). The beautiful Franklin tree (see Figure 2.1) grows only in cultivation and is no longer found in the wild. The long-term goal of many ex situ conservation programs is the eventual establishment of new populations in the wild, once sufficient number of individuals and a suitable habitat are available.

Ex situ facilities for animal preservation include zoos, game farms, aquariums, and captive breeding programs. Plants are maintained in botanical gardens, arboretums, and seed banks. An intermediate strategy that combines elements of both ex situ and in situ preservation is the intensive monitoring and management of populations of rare and endangered species in small protected areas; such populations are still somewhat wild, but human intervention may be used on occasion to prevent population decline.

Ex situ conservation efforts are an important part of an integrated conservation strategy to protect endangered species (Falk

3.19 Père David's deer (*Elaphurus davidianus*) has been extinct in the wild since about 1200 B.C. The species remained only in managed hunting reserves kept by Chinese royalty, and is now kept in captive herds. (Photograph by Jessie Cohen, National Zoological Park, Smithsonian Institution.)

3.20 Modern zoos offer educational opportunities to the public as well as serving as sanctuaries for animals. These visitors to the Bronx Zoo are observing prairie dogs in a surrounding that lets them imitate the animals' behavior. (Photograph by Michael K. Nichols/National Geographic Image Collection.)

1991). Ex situ and in situ conservation strategies are complementary approaches (Robinson 1992). Individuals from ex situ populations can be periodically released into the wild to augment in situ conservation efforts. Research on captive populations can provide insight into the basic biology of a species and suggest new conservation strategies for in situ populations. Ex situ populations that are self-maintaining can also reduce the need to collect individuals from the wild for display or research purposes. Finally, displaying individuals from endangered species can help to educate the public about the need to preserve the species, and so protect other members of the species in the wild (Figure 3.20). In situ preservation of species, in turn, is vital to the survival of species that are difficult to maintain in captivity, such as the rhinoceros, as well as to the continued ability of zoos, aquariums, and botanical gardens to display new species. Ex situ conservation is not cheap; the cost of maintaining African elephants and black rhinos in zoos is 50 times greater than protecting the same number of individuals in East African national parks (Leader-Williams 1990); the cost of maintaining U.S. zoos is around $1 billion per year. However, as Michael Soulé says, "There are no hopeless cases, only people without hope and expensive cases" (Soulé 1987).

Zoos

Zoos, along with affiliated universities, government wildlife departments, and conservation organizations, presently maintain over

700,000 individuals representing 3000 species of mammals, birds, reptiles, and amphibians (WCMC 1992). While this number of captive animals may seem impressive, it is trivial in comparison to the numbers of domestic cats, dogs, and fish kept by people as pets. The emphasis zoos place on displaying "charismatic megafauna," such as pandas, giraffes, and elephants, tends to ignore the enormous threats to the huge numbers of insects and other invertebrates that form the majority of the world's animal species.

A current goal of most major zoos is to establish captive breeding populations of rare and endangered animals. Only a small proportion of rare mammal species kept by zoos worldwide currently have self-sustaining captive populations of sufficient size to maintain their genetic variation (Ralls and Ballou 1983; WCMC 1992). To remedy this situation, zoos and affiliated conservation organizations have embarked on a major effort to build the facilities and develop the technology necessary to establish breeding colonies of rare and endangered animals, such as the snow leopard and the orangutan, and to develop new methods and programs for reestablishing species in the wild (Figure 3.21). Some of these facilities are highly specialized, such as the International Crane Foundation in Wisconsin, which is attempting to establish captive breeding populations of all crane species.

Ex situ conservation efforts have been increasingly directed at saving endangered species of invertebrates as well, including butterflies, beetles, dragonflies, spiders, and mollusks. This is important because there are far more species of invertebrate than vertebrates, and many invertebrate species are restricted in distribution and declining in numbers. Other important targets for ex situ conservation efforts are rare breeds of domestic animals on which hu-

3.21 Snow leopards (*Panthera uncia*) reproduce well in captivity. Maintaining breeding colonies of these animals can reduce zoos' need to capture individuals from the declining wild population. Since 1974, the majority of snow leopards in zoos have been born in captivity (white bars), and fewer animals have been caught in the wild (shaded bars). (After Blomqvist 1995.)

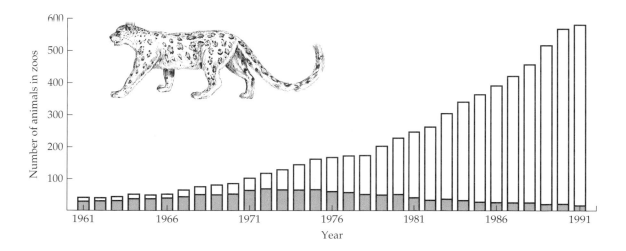

man societies depend for animal protein, dairy products, leather, wool, agricultural labor, transport, and recreation (Hall and Ruane 1993). Secure populations of these breeds are a critical genetic resource for the improvement and long-term health of our supplies of pigs, cattle, chickens, sheep, and other domestic animals.

The success of captive breeding programs has been enhanced by several programs that collect and disseminate knowledge about rare and endangered species. The Species Survival Commission Conservation Breeding Specialist Group, a division of the IUCN (the International Union for the Conservation of Nature), and affiliated organizations, such as the American Zoo and Aquarium Association, provide zoos with the information necessary for proper care and handling of these species, including data on nutritional requirements, anaesthetic techniques, optimal housing conditions, and vaccinations and antibiotics. Central databases of breeding records and stud books are being developed to prevent matings among close relatives and the resulting higher offspring mortality associated with genetic drift and inbreeding depression. One of the most important of these databases is the International Species Inventory System (ISIS), which provides information on 4200 kinds of animals at 395 zoological institutions in 39 countries.

A wide range of innovative techniques are being developed to increase the reproductive rates of captive species. Some of these come directly from human and veterinary medicine, while others are novel methods developed for particular species (Kleiman et al. 1996). These techniques include **cross-fostering**, having mothers of common species raise the offspring of rare species; **artificial insemination** when adults show no interest in mating or are living in different locations; **artificial incubation** of eggs in ideal hatching conditions; and **embryo transfer**, the implanting of fertilized eggs of rare species into surrogate mothers of common species (Figure 3.22). One novel approach involves freezing the eggs, sperm, embryos, and tissues of species on the verge of extinction—so called "frozen zoos." The hope is that in the future, new techniques, such as cell cloning, can be used to recreate these species.

When scientists decide to use these methods to preserve a species, they must address a series of ethical questions (Norton et al. 1995). First, how necessary and how effective are these methods for a particular species? Is it better to let the last few individuals of a species live out their days in the wild than to start a captive population that may be unable to readapt to wild conditions? Second, does a population of a rare species that has been raised in captivity and does not know how to survive in its own natural environment really represent survival for the species? Third, are species held in captivity for their own benefit or for the benefit of zoos?

3.22 This bongo calf (*Tragelaphus euryceros*), an endangered species, was produced by embryo transfer using an eland (*Taurotragus oryxii*) as a surrogate mother at the Cincinnati Zoo Center for Reproduction of Endangered Wildlife. (Photograph © Cincinnati Zoo.)

Even when the answers to these questions indicate that ex situ management is appropriate, it is not always feasible to create ex situ populations of rare animal species. A species may have been so severely reduced in numbers that it has low breeding success and high infant mortality due to inbreeding depression. Certain animals, particularly marine mammals, are so large or require such specialized environments that the facilities for maintaining and handling them are prohibitively expensive. Many invertebrates have complex life cycles in which their diets change as they grow and in which their environmental needs vary in subtle ways. Many of these species are not possible to raise with our present knowledge. Finally, certain species are simply difficult to breed, despite the best efforts of scientists. Two prime examples are the giant panda and the Sumatran rhino, which have low reproductive rates in the wild and have not reproduced well in captivity despite a considerable effort to find effective breeding methods (Schaller 1993).

Aquariums

To deal with the threats to aquatic species, ichthyologists, marine biologists, and coral reef experts who work for public aquariums are increasingly working cooperatively with colleagues in marine research institutes, government fisheries departments, and conservation organizations to develop programs for the conservation of rich natural aquatic communities and species of special concern. At present, approximately 600,000 individual fish are maintained in aquariums, with most of these collected from the wild (Olney and Ellis 1991). Major efforts presently are being made to develop breeding techniques to maintain rare species in aquariums, sometimes for release back into the wild, and also to eliminate the need to collect wild specimens (Philippart 1995). Many of the techniques used in fish breeding were originally developed by fisheries biologists for large-scale stocking operations involving trout, bass, salmon, and other commercial species. Other techniques were discovered in the aquarium pet trade as dealers attempted to propagate tropical fish for sale. These techniques are now being applied to such endangered freshwater fauna as the desert pupfishes of the American Southwest, stream fishes of the Tennessee River Basin, and cichlids of the African Rift lakes. Programs for breeding endangered marine fishes and coral species are still in an early stage, but this is an area of active research at the present time. Also, as aquaculture increasingly supplies human populations with fish, mollusks, and shrimp, breeding programs are being developed for the genetic stocks needed to improve these species and protect them against disease and unforeseen threats.

Aquariums have a particularly important role to play in the conservation of endangered cetaceans. Aquarium personnel often respond to public requests for assistance in dealing with whales stranded on beaches or disoriented in shallow waters. The aquarium community potentially can use the lessons learned from working with common captive species, such as the bottle-nosed dolphin, to develop programs to aid endangered species (Figure 3.23).

Botanical gardens and arboretums

The world's 1600 botanical gardens contain major collections of living plants and represent a crucial resource for plant conservation efforts. The botanical gardens of the world are currently growing 4 million plants, representing 80,000 species, approximately 30% of the world's flora. Additional species are being grown in greenhouses, subsistence gardens, hobby gardens, and other such situa-

3.23 The breeding of bottle-nosed dolphins (*Tursiops truncatus*) in captivity has provided aquarium personnel with valuable experience that can be applied to endangered cetacean species. Shown here are a mother and calf. (Photograph courtesy of Sea World.)

tions (though often with very few individuals per species). The world's largest botanical garden, the Royal Botanic Gardens, Kew, in England, has an estimated 25,000 plant species under cultivation—about 10% of the world's total—of which 2700 are endangered or threatened (Figure 3.24). Botanical gardens need to increase the number of individuals grown for every species in order to protect the range of genetic variability found in each.

Botanical gardens increasingly focus on the cultivation of rare and endangered plant species, and many specialize in particular types of plants. The Arnold Arboretum of Harvard University grows hundreds of different temperate tree species. The New England Wild Flower Society has a collection of hundreds of perennial temperate herbs at its Garden in the Woods. In California, a specialized pine arboretum grows 72 of the world's 110 species of pines, while South Africa's leading botanical garden has 25% of that country's many plant species growing under cultivation.

Botanical gardens are in a unique position to contribute to conservation efforts because their living collections and their associated herbaria of dried plant collections represent one of the best sources of information on plant distribution and habitat requirements (Given 1994). The staff of botanical gardens are often recognized authorities on plant identification and conservation status. Expeditions sent out by botanical gardens discover new species and make observations on known species, while over 250 botanical gardens maintain nature reserves that serve as important conservation areas in their own right. In addition, botanical gardens are in a position to

3.24 The Royal Botanic Gardens, Kew, is well-known for its training courses and research in plant conservation and horticulture. Here a training session is being conducted around a collection of desert plants inside the Princess of Wales Conservatory. (Photograph courtesy of the Royal Botanic Gardens, Kew.)

educate the public about conservation issues because an estimated 150 million people per year visit them.

At the international level, the Botanical Gardens Conservation Secretariat (BGCS) of The International Union for the Conservation of Nature is organizing and coordinating conservation efforts by the world's botanical gardens. The priorities of this program include the development of a worldwide database system for coordinating collecting activity and identifying important species that are under-represented or absent from living collections. A problem with the distribution of botanical gardens is that most of them are located in the temperate zone, even though most of the world's plant species are found in the Tropics. While a number of major gardens exist in such places as Singapore, Sri Lanka, Java, and Colombia, establishing new botanical gardens in the Tropics should be a priority for the international conservation community, along with training local plant taxonomists to fill staff positions.

Seed banks. In addition to growing plants, botanical gardens and research institutes have developed collections of seeds, sometimes known as **seed banks**, collected from the wild and from cultivated plants. When seeds are collected, efforts are made to include the range of genetic variation found in the species. This is done by collecting seeds from populations growing across the range of the

(A)

(B)

(C)

(D)

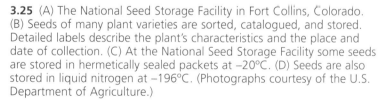

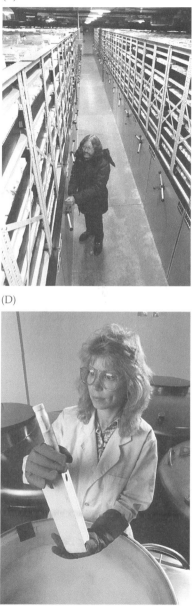

3.25 (A) The National Seed Storage Facility in Fort Collins, Colorado. (B) Seeds of many plant varieties are sorted, catalogued, and stored. Detailed labels describe the plant's characteristics and the place and date of collection. (C) At the National Seed Storage Facility some seeds are stored in hermetically sealed packets at −20°C. (D) Seeds are also stored in liquid nitrogen at −196°C. (Photographs courtesy of the U.S. Department of Agriculture.)

species, sampling seeds from populations growing in different environmental conditions, and collecting seeds from many individuals in each population. The seeds of most plant species can be stored in cold, dry conditions in these seed banks for long periods of time and then later germinated (Figure 3.25). The ability of seeds to remain dormant is extremely valuable for ex situ conservation efforts because it allows the seeds of large numbers of rare species to be frozen and stored in a small space, with minimal supervision, and at a low cost. More than 50 major seed banks exist in the world, many of them in developing countries, with their activities coordinated by the Consultative Group on International Agricultural Research (CGIAR) (Rhoades 1991; Fuccilo et al. 1998).

Even though seed banks have great potential for conserving rare and endangered species, they have certain problems as well. If power supplies fail or equipment breaks down, the entire frozen collection may be damaged. Even in cold storage, seeds gradually lose their ability to germinate due to exhaustion of their energetic reserves and the accumulation of harmful mutations. To overcome this gradual deterioration of seed quality, seed samples must be periodically germinated, adult plants grown to maturity, and new seed samples stored. For seed banks with large collections, this testing and rejuvenation of seed samples can be a formidable task.

Approximately 15% of the world's plant species have "recalcitrant" seeds that either lack dormancy or do not tolerate low-temperature storage conditions, and consequently cannot be stored in seed banks. Seeds of these species must germinate right away or die. Species with recalcitrant seeds are much more common in the tropical forest than in the temperate zone; the seeds of many economically important tropical fruit trees, timber trees, and plantation crops, such as cocoa and rubber, cannot be stored. Some of these plant species can be maintained in tissue culture in controlled conditions or propagated by cuttings from a parent plant, though these processes are currently more expensive than growing plants from seeds.

Seed banks have been embraced by the international agricultural community as an effective way of preserving the genetic variability that exists in agricultural crops. Often genes for resistance to a particular disease or pest are found in only one variety of a crop, known as a **land race**, which is grown in only one small area of the world (Figure 3.26). This genetic variability is often crucial to the agricultural industry in its efforts to maintain and increase the high productivity of modern crops and to respond to changing environmental conditions, such as acid rain, drought, and soil salinity. Researchers are in a race against time to preserve this genetic variability because traditional farmers throughout the world are abandoning their local crop varieties in favor of standard, high-yielding varieties (Altieri and Anderson 1992; Cleveland et al. 1994). This worldwide phenomenon is illustrated by Sri Lankan farmers, who grew 2000 different varieties of rice until the late 1950s, when they switched over to just 5 high-yielding varieties. So far, over 2 million collections of seeds have been acquired by agricultural seed banks. Many of the major food crops, such as wheat, corn (maize), oats, and potatoes,

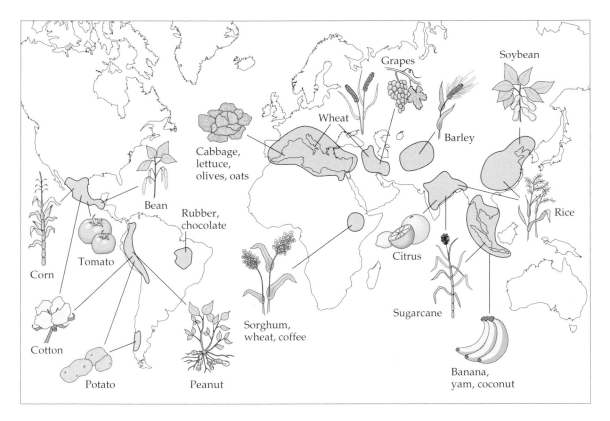

3.26 Crop species show high genetic diversity in certain areas of the world, often where the species was first domesticated or where it is still grown in traditional agricultural settings. (Courtesy of Garrison Wilkes.)

are well represented in seed banks, and other important crops such as rice, millet, and sorghum are being intensively collected as well. However, crops of only regional significance, medicinal plants, fiber plants, and other useful plants are not as well represented in seed banks. Wild relatives of crop plants are also not adequately represented in seed banks even though these species are extremely useful in crop improvement programs.

Special efforts are being made to protect the genetic resources of commercially important tree species (Ledig 1988; Rogers and Ledig 1996). Storage of seeds is difficult for many important genera of trees, such as oaks (*Quercus*) and poplars (*Populus*). Even pine seeds cannot be stored indefinitely and must eventually be grown out as trees. In many cases, seeds from selected trees are used to establish seed orchards for producing commercial seeds. Preservation of natural areas where commercial species occur is increasingly being considered as a way of protecting the genetic variability needed for forestry. International cooperation is needed in forestry research and conservation because commercial species are often grown far from their countries of origin; for example, Monterey pine (*Pinus radiata*) from the United States is planted on 3 million ha distributed throughout Chile, New Zealand, Australia, and Spain.

A major controversy over the development of seed banks is who owns and controls the genetic resources of crop plants (Brush and Stabinsky 1996). The genes of local land races of crop plants and wild relatives of crop species provide the building blocks needed to develop advanced "elite" high-yielding varieties suitable for modern agriculture. An estimated 96% of the genetic variability that is necessary for modern agriculture comes from the developing countries of the world, such as India, Ethiopia, Peru, Mexico, Indonesia, and Egypt, yet the breeding programs for "elite" strains frequently take place in the industrialized countries of North America and Europe. In the past, international seed banks freely collected seeds and plant tissue from developing countries and gave them to research stations and seed companies. Yet, once seed companies developed new "elite" strains through sophisticated breeding programs and field trials, they sold their seeds at a high price to maximize profits. Developing countries are now questioning why they should share their genetic resources freely but then have to pay for advanced seeds based on those genetic resources. One solution to this dilemma involves negotiating agreements using the framework of the Convention on Biological Diversity (see Chapter 5), in which countries agree to share their genetic resources in exchange for receiving new products and a share of the profits (Vogel 1994). Such agreements would include provisions for protecting biological diversity.

Conservation Categories of Species

To highlight the status of rare species for conservation purposes, IUCN has established 10 conservation categories (IUCN, 1996); species in categories 2–4 are considered to be "threatened" with extinction. These categories have proved useful at the national and international levels by directing attention toward species of special concern and by identifying species threatened with extinction for protection through international agreements such as the Convention on International Trade in Endangered Species (CITES).

1. *Extinct*: A species (or other taxa, such as subspecies and varieties) that is no longer known to exist. Exhaustive and repeated searches of localities where the species was once found, and of other possible sites, have failed to detect these species.

2. *Extinct in the wild*: The species exists only in cultivation, in captivity, or as naturalized populations well outside its original range. Searches of its known localities have failed to detect the species.

3. *Critically endangered*: Species that have an extremely high risk of going extinct in the wild in the immediate future. Of particular concern are species whose numbers of individuals have been declining and are reduced to the point where survival of the species is unlikely if present trends continue.

4. *Endangered*: Species that have a high risk of extinction in the wild in the near future, and may become critically endangered.

5. *Vulnerable*: Species that have a high risk of extinction in the wild in the medium-term future, and may become endangered.

6. *Conservation dependent:* The species is not currently threatened, but is dependent on a conservation program, without which the species would be threatened with extinction.

7. *Near threatened*: The species is close to qualifying as vulnerable, but is not currently considered threatened.

8. *Least concern:* The species is not considered near threatened or threatened.

9. *Data deficient:* Inadequate information exists to determine the risk of extinction for the species. In many cases, species have not been seen for years or decades because no biologists have made the effort to look for them. More information is required before the species can be assigned to a threat category.

10. *Not evaluated*: The species has not yet been assessed for its threat category.

Given the legal restrictions that accompany these assignments, and the resulting financial implications for landowners, corporations, and governments, the definitions of each category needed to be clarified to prevent arguments. To clarify this situation, in 1994 the IUCN issued refined and more quantitative definitions and guidelines for categorization in a three-level system of classification based on the probability of extinction (Mace and Lande 1991; IUCN 1994b, 1996):

1. *Critically endangered species* have a 50% or greater probability of extinction within 10 years or 3 generations, whichever is longer.

2. *Endangered* species have a 20% probability of extinction within 20 years or 5 generations.

3. *Vulnerable* species have a 10% or greater probability of extinction within 100 years.

Assignment of categories depends on having at least one of the following types of information:

1. Observable decline in numbers of individuals.

2. The size of the geographical area occupied by the species, and the number of populations.

3. The total number of individuals alive, and the number of breeding individuals.

4. The expected decline in the numbers of individuals if current and projected trends in populations decline or habitat destruction continue.

5. The probability of the species going extinct in a certain number of years or generations.

These quantitative criteria for assigning categories are based on the developing methods of population viability analysis and focus particularly on population trends and habitat condition. For example, a critical species has at least one of the following characteristics: total population size less than 250 individuals or less than 50 breeding individuals; population has declined by 80% or more over the last 10 years or generations; a more than 25% decline in population numbers is expected within 3 years or one generation; or the overall extinction probability is greater than 50% in 10 years or 3 generations. Species can also be assigned critical status as a result of restricted range of the species (less than 100 km^2 at a single location), observed or predicted habitat loss, ecological imbalance, or commercial exploitation (Figure 3.27). The use of habitat loss in

3.27 Yellow gentian (*Gentiana lutea*), a beautiful perennial herb of European mountain meadows, has roots that are collected for traditional medicine. Approximately 1500 tons of dried roots are used each year in a wide variety of preparations to stimulate digestion and to treat stomachache. Due to overharvesting and the resulting decline and destruction of many populations, the species is listed as endangered in Portugal, Albania, and certain regions of Germany and Switzerland, and as vulnerable in other countries, according to the IUCN's classification categories. Despite official regulation that restricts collection to designated areas, illegal harvesting continues. (Photograph by Bob Gibbons, Natural Image.)

TABLE 3.1 Percentage of species in some temperate countries that are threatened with global extinction[a]

	Mammals		Breeding birds		Reptiles		Amphibians		Plants	
	No.[b]	%	No.	%	No.	%	No.	%	No.	%
Argentina	320	8.4	897	4.6	220	2.3	145	3.4	9000	1.9
Canada	193	3.6	426	1.2	41	7.3	41	2.4	2920	22.2
China	394	19.0	1100	8.2	340	4.4	263	0.4	30,000	1.1
Japan	132	22.0	>250	13.2	66	12.1	52	19.2	4700	15.0
Russia	269	11.5	628	6.1	58	8.6	23	0.0	—	—
South Africa	247	13.4	596	2.7	299	6.4	95	9.5	23,000	4.1
United Kingdom	50	8.0	230	0.9	8	0.0	7	0.0	1550	1.8
United States	428	8.2	650	7.7	280	10.0	233	10.3	16,302	11.3

Source: Data from WRI 1998.
[a]Threatened species include those in the IUCN categories "critically endangered," "endangered," and "vulnerable."
[b]Number of species.

assigning categories is particularly useful for species that are poorly known biologically, such as many tropical insect species; species can be listed as threatened if their habitat is being destroyed.

The advantage of this system is that it provides a standard, quantitative method of classification by which decisions can be reviewed and evaluated by other scientists, according to accepted quantitative criteria and using whatever information is available. However, this method could be applied arbitrarily, if decisions have to be made with insufficient data; gathering the data needed for this approach could be too expensive and time-consuming, particularly in developing countries and in rapidly changing situations. Regardless of these limitations, the new system of species classification is a distinct improvement and will assist attempts to protect species.

Using the IUCN categories, the World Conservation Monitoring Centre (WCMC) has evaluated and described the threats to about 60,000 plant and 5000 animal species in its series of Red Data Books (IUCN 1990, 1996). The great majority of the species on these lists are plants, reflecting the trend of listing plant species in threatened habitats. However, there are also numerous listed species of fish (700), amphibians (100), reptiles (200), mollusks (900), insects (500), inland water crustaceans (400), birds (1100), and mammals (1100). The IUCN system has been applied to specific geographical areas and groups of species as a way of highlighting conservation priorities. As a group, mammals face a greater degree of threat than birds; comparing regions, in general, the species of Japan are generally more threatened than the species of South Africa, which are more threatened, in turn, than the species of the United Kingdom (Table 3.1). Malaysia provides a detailed example (Kiew 1991):

- Of 2830 tree species in peninsular Malaysia, 511 species are considered threatened.

- A large number of Malaysian herb species are endemic to single localities, such as mountaintops, streams, waterfalls, or limestone outcrops. These species are threatened with extinction if their habitat is destroyed.

- All five species of sea turtles in Malaysia are considered endangered due to a combination of habitat loss, egg collecting, hunting, pollution of marine waters, unregulated tourism, and entanglement in fishing nets.

- Over 80% of Malaysian Borneo's primate species are under some threat, due to a combination of habitat destruction and hunting.

A program similar to the efforts of the IUCN and WCMC is the network of Natural Heritage Data Centers that covers all 50 of the United States, 3 provinces in Canada, and 14 Latin American countries (Jenkins 1996). This program, strongly supported by The Nature Conservancy, gathers, organizes, and manages information on the occurrence of what it refers to as "elements of conservation interest": more than 35,000 species and 7000 subspecies, as well as biological communities. Elements are given status ranks based on a series of standard criteria: the number of remaining populations or occurrences, number of individuals remaining (for species) or aerial extent (for communities), number of protected sites, degree of threat, and innate vulnerability of the species or community. The Nature Conservancy approach has been applied in detail to the species of the United States. The results, given by Stein and Flack in the *Species Report Card: The State of U.S. Plants and Animals* (1997), demonstrate that aquatic species groups, including freshwater mussels, crayfish, amphibians, and fish, are in greater danger of extinction than other well-known terrestrial groups, such as insects, birds, and mammals (Figure 3.28). Freshwater mussels are by far the most endangered species group, with 11.8% of these species presumed to be extinct already and almost 25% critically imperiled. Land plants are intermediate in degree of endangerment.

This system has proved to be extremely successful and useful in organizing between 300,000 to 400,000 records of occurrence. Regional data centers are maintained by hundreds of staff and are consulted approximately 200,000 times a year for information to assist protection efforts on behalf of endangered species, environmental impact reports, scientific research, and land use decisions. Organizing vast amounts of conservation information is an expensive, labor-intensive activity, but it is a crucial component of conservation efforts. We need to know what species and biological communities are in danger and where they occur in order to protect them.

In Switzerland, efforts are underway to evaluate the conservation status of the 2106 species of plants and animals that are cur-

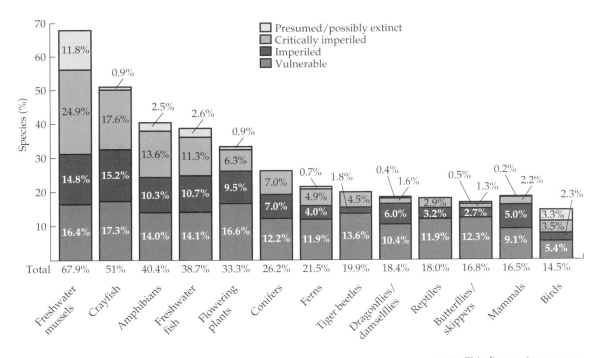

3.28 This figure shows some species groups studied in the United States and ranked, according to criteria endorsed by The Nature Conservancy, as critically imperiled, imperiled, or vulnerable (rankings 1–3, respectively, on a scale of 5). The graph also shows the percentage of species in each class that are presumed to be extinct. (From Stein and Flack 1997.)

rently on the country's Red Lists of threatened species. Of these, 317 species have been identified as stable or increasing in abundance, thanks to conservation and protection measures; these 317 now form a "Blue List" of recovering species that highlights conservation success (Gigon et al. 1998).

Legal Protection of Species

After species have been identified as needing protection, laws can be passed and treaties can be signed to implement this effort. National laws protect species within individual countries; international agreements regulate the trade of species between countries.

National laws

Throughout much of the modern world, national governments and national conservation organizations play a leading role in the protection of all levels of biological diversity. Laws are passed establishing national parks, regulating activities such as fishing, logging, and grazing, and limiting air and water pollution. International treaties affecting trade in endangered animals are implemented at the national level and enforced at borders. Many national laws are targeted toward preserving species. The effectiveness with which these laws are enforced shows a nation's determination to protect its natural resources and its citizens. In many countries, it is being recognized that preserving a healthy environment and protecting species are linked to the health of the people.

Countries differ in their approaches to protecting biological diversity. The 15 member states of the European Union, for example, rely on international conventions, such as the Convention on International Trade in Endangered Species and the Convention on Biological Diversity, to protect species. In addition, they have enacted specific regulations and directives to protect biological diversity. One example is the Bird Directive, requiring member countries to protect and manage habitats for birds, particularly those that involve migratory and breeding species (McLean et al. 1999). Implementation and enforcement of these conservation measures varies widely among the European countries.

In the United States, the principal law protecting species is the Endangered Species Act of 1973 (Endangered Species Coalition 1992), with additional coverage provided by the Marine Mammals Protection Act. The U.S. Endangered Species Act was created by the Congress to "provide a means whereby the ecosystems upon which endangered species and threatened species depend may be conserved and to provide a program for the conservation of such species." Species are protected under the Act if they are placed on an official list of endangered and threatened species. As defined by the law, endangered species are those that are likely to become extinct, as a result of human activities or natural causes, in all or a major portion of their ranges; threatened species are those likely to become endangered in the near future. The Secretary of the Interior, acting through the U.S. Fish and Wildlife Service, and the Secretary of Commerce, acting through the National Marine Fisheries Service (NMFS), can add and remove species from the list based on information available to them. In addition, a recovery plan is required for each listed species, typically involving habitat preservation, habitat restoration, and active management of the species (Foin et al. 1998). More than 1100 species in the United States have been placed on the list, as well as about 550 species from elsewhere in the world. At the present time almost 4000 species are candidates for listing; while awaiting official listed status, some of these species have probably become extinct. The Act requires all U.S. government agencies to consult with the Fish and Wildlife Service and the NMFS to determine whether their activities will affect listed species, and prohibits activities that will harm those species or their habitat. The law also prevents private individuals, businesses, and local governments from harming or "taking" listed species, and prohibits all trade in listed species.

Since its enactment more than two decades ago, the Endangered Species Act has become increasingly important as a conservation tool. The Act has provided the legal basis for protecting some of the most significant animal species in the United States, such as the grizzly bear, the bald eagle, the whooping crane, and the gray wolf. Because the law protects the ecosystems in which endangered species live, entire biological communities and thousands of other species have in effect been protected at the same time (Carroll et al. 1996). Although it has brought about the protection of thousands of species, an analysis of the Act reveals a number of troubling tendencies. The list of species it safeguards is not as comprehensive as it probably should be, nor does it necessarily reflect those species or groups of species that are most at risk of extinction. The great majority of species listed under the Act are plants and vertebrates, despite the fact that most species are insects and other invertebrates. For instance, about half of the 300 freshwater mussel species found in the United States are declining, in danger of extinction, or are already extinct, yet only 56 species are listed under the Act (Stolzenburg 1992; Chadwick 1995). Clearly, greater efforts must be made to study the various invertebrate groups and extend listing to endangered species whenever necessary. Another study of species covered by the Act has shown that animal species typically have only about 1000 remaining individuals at the time of listing, and plants species often have fewer than 120 individuals remaining (Wilcove et al. 1993). Populations this small may encounter the genetic and demographic problems associated with small population size that can prevent recovery. At the extreme were 39 species listed when they had 10 or fewer individuals remaining, and a freshwater mussel that was listed when it had only a single remaining population that was not reproducing. Endangered species probably should be given protection under the Act before they decline to the point at which recovery becomes difficult. An earlier listing of a declining species might allow it to recover and become a candidate for de-listing more quickly.

While the Endangered Species Act has been a model for other countries, such as Australia, its implementation in this country has often been controversial (Chadwick 1995; Easter-Pilcher 1996). Ever since its enactment, this legislation has been a source of contention between conservation efforts and business interests in the United States. The protection afforded to listed species is so strong that business interests often lobby vigorously against the listing of species in their area. One reason that business leaders are reluctant to allow new species to be added to the list is the difficulty of rehabilitating species to the point at which they can be removed from the list (Tear et al. 1993). So far only 21 of the listed species have been taken off the list, the most notable successes being the brown pelican and the American alligator. In 1994 the bald eagle was moved from the highly regulated "endangered" category to the less critical "threatened" category, in recognition that its numbers had increased from 400 breeding pairs in the 1960s to the current 4000. The difficulty of implementing recovery plans is often not primarily biological, but to a large extent political, administrative, and ultimately, financial. The U.S. Fish and Wildlife Service annually spends around $50 million on activities related to the Act, but one estimate suggests that over $4 billion is needed to remove the threat. Though funding for species recovery has been increasing, the number of listed species has been growing even faster, so there is less money available per species. One solution might be to allow private landowners to receive tax deductions for maintaining habitats for endangered species.

In an attempt to find compromises between the economic interests of the country and conservation priorities, the Endangered Species Act was amended in 1978 to allow a Cabinet-level committee, the so-called "God Squad," to exclude areas from protection. The clash of interests is illustrated dramatically by the controversy over the designation of 2.8 million ha of old-growth forest in the Pacific Northwest, potentially worth billions of dollars, as critical protected habitat for the northern spotted owl. Limitations on logging in this region, strongly advocated by environmental organizations, have been fiercely resisted by business and citizen groups as well as by many politicians. After years of negotiations, legal maneuvering, and political lobbying, a solution to this ongoing controversy still has not been found. Recognition that intact watersheds are needed to maintain salmon populations and the valuable commercial and sport fisheries they support may eventually tip the balance toward the preservation of these forests.

Concerns about the implications of the Endangered Species Act have often forced business organizations, conservation groups, and governments to develop compromise **Habitat Conservation Plans** that reconcile conservation and business interests (Hoffman et al. 1997; Noss et al. 1997). One such plan involves the coastal sage scrub community of Southern California that includes 100 rare, sensitive, threatened, or endangered species, including the California gnat-catcher (*Polioptila californica californica*). After fighting each other for years over this valuable real estate, developers, farmers, environmental groups, and governmental agencies have agreed to a compromise that protects the most important natural communities, but still allows limited development on some lower-quality sites (Figure 3.29). The plan covers 160,000 ha of habitat including 50 cities, 5 counties, and several military bases. While the plan is not perfect, it is at least an attempt to create the next generation of conservation planning: a multispecies, ecosystem- or community-based approach extending over a wide geographical region that includes many projects, landowners, and jurisdictions. In this case and others, the

(A)

3.29 (A) In southern California, a Habitat Conservation Plan has been established to protect the California gnatcatcher, shown here at a nest with chicks. (Photograph by Robb Hirsch.) (B) Protecting large blocks of coastal sage scrub community from uncontrolled development and fragmentation is key to the plan. (Photograph by Reed Noss).

(B)

result is a compromise in which economic activities proceed but pay a higher cost to support conservation activities.

International agreements

Although the major mechanisms protecting biological diversity throughout the world today are based within individual countries, agreements at the international level are increasingly being used to protect species and habitats. International cooperation is an absolute requirement for several crucial reasons. First, species often migrate across international borders. Conservation efforts to protect migratory bird species in northern Europe will not work if the birds' overwintering habitat in Africa is destroyed.

Second, international trade in biological products can result in the overexploitation of species to supply the demand. Control and management of the trade are required at the points of both export and import.

Third, the benefits of biological diversity are of international importance. Wealthy countries of the temperate zone that benefit from tropical biological diversity need to be willing to help the less wealthy countries of the world that preserve it.

Finally, many of the problems that threaten species and ecosystems are international in scope and require international cooperation to solve. Such threats include overfishing and overhunting, atmospheric pollution and acid rain, pollution of lakes, rivers, and oceans, global climate change, and ozone depletion.

The single most important treaty protecting species at an international level is the Convention on International Trade in Endangered Species (CITES), established in 1973 in association with the United Nations Environmental Programme (UNEP) (Wijnstekers 1992; Hemley 1994). The treaty is currently endorsed by more than 120 countries. CITES establishes lists of species whose international trade is to be controlled; the member countries agree to restrict trade in and destructive exploitation of those species. Appendix I of the treaty includes approximately 675 animals and plants whose commercial trade is prohibited, and Appendix II includes about 3700 animals and 21,000 plants whose international trade is regulated and monitored. Among plants, Appendixes I and II cover such important horticultural species as orchids, cycads, cacti, carnivorous plants, and tree ferns; increasingly they cover timber species as well. Among animals, closely regulated groups include parrots, large cat species, whales, sea turtles, birds of prey, rhinoceroses, bears, primates, species collected for the pet, zoo, and aquarium trades, and species harvested for their fur, skin, or other commercial products.

International treaties such as CITES are implemented when a country signing the treaties passes laws making it a criminal act to violate them. Once CITES laws are passed within a country, police, customs inspectors, wildlife officers, and other government agents can arrest and prosecute individuals possessing or trading in CITES-listed species and seize the products or organisms involved (Figure 3.30). Technical advice to countries on how to implement the treaty is provided by such nongovernmental organizations as the International Union for the Conservation of Nature Wildlife Trade Specialist Group, the World Wildlife Fund (WWF) TRAFFIC Network, and the World Conservation Monitoring Centre (WCMC) Wildlife Trade Monitoring Unit. The CITES treaty's most notable success has been the 1989 ban on the ivory trade, which was causing severe declines in African elephant populations (Poole 1996).

3.30 In August of 1995, Belgian customs inspectors seized an enormous shipment of contraband wildlife items. The haul contained many items banned by the CITES treaty, including monkey skulls, stuffed specimens of rare bird species, and tiger pelts. Parts from over 2000 individual animals were identified in the shipment. (AP/Wide World Photos.)

Another international treaty is the Convention on Conservation of Migratory Species of Wild Animals, signed in 1979, which focuses primarily on bird species. This convention serves as an important complement to CITES by encouraging international efforts to conserve bird species that migrate across international borders and by emphasizing regional approaches to research, management, and hunting regulations. However, there are problems with this convention: only 36 countries have signed it and its budget is very limited. It also does not cover other migratory species, such as marine mammals and fish.

Other important international agreements that protect species include

- The Convention on Conservation of Antarctic Marine Living Resources
- The International Convention for the Regulation of Whaling, which established the International Whaling Commission
- The International Convention for the Protection of Birds and the Benelux Convention on the Hunting and Protection of Birds
- The Convention on Fishing and Conservation of Living Resources in the Baltic Sea
- The International Commission on Atlantic Tuna
- Miscellaneous regional agreements protecting specific groups of animals, such as prawns, lobsters, crabs, fur seals, salmon, bats, and vicuña

A weakness of these international treaties is that participation is voluntary; countries can withdraw from the convention to pursue their own interests when they find the conditions of compliance too difficult (Young 1999). This flaw was highlighted when several countries decided simply to disregard the International Whaling Commission because of its ban on hunting. Persuasion and public pressure are necessary to induce countries to enforce the provisions of the treaties and prosecute violators.

Summary

1. Biologists have observed that small populations have a greater tendency to go extinct than large populations. The minimum viable population size (MVP) is the number of individuals necessary to ensure that a population has a high probability of surviving for the foreseeable future.

2. Small populations are subject to rapid extinction for three main reasons: loss of genetic variability and inbreeding depression, demographic fluctuations, and environmental variation including natural catastrophes. The combined effects of these factors have been compared to a vortex that tends to drive small populations to extinction. Population viability analysis uses demographic, genetic, environmental, and natural catastrophe data to estimate a population's MVP and its probability of persistence in an environment.

3. Conservation biologists often determine whether an endangered species is stable, increasing, fluctuating, or declining by monitoring its populations. Often the key to protecting and managing a rare or endangered species is an understanding of its natural history. Some rare species are more accurately described as metapopulations in which a mosaic of populations is linked by some degree of migration and recolonization.

4. New populations of rare and endangered species can be established in the wild using either captive-raised or wild-caught individuals. Mammals and birds raised in captivity may require social and behavioral training before release, and they often require some degree of maintenance after release. Reintroduction of plant species requires a different approach because of their specialized environmental requirements at the seed and seedling stages.

5. Some species that are in danger of going extinct in the wild can be maintained in zoos, aquariums, and botanical gardens; this strategy is known as ex situ conservation. These captive colonies can sometimes be used later to reestablish species in the wild.

6. To highlight the status of species for conservation purposes, the IUCN has established conservation categories: extinct, extinct in the wild, critically endangered, endangered, vulnerable, conservation dependent, near threatened, least concern, data deficient, and not evaluated. This system of classification, which is increasingly based on quantitative evaluation of populations, is now widely used to evaluate the status of species and establish conservation priorities.

7. One of the most effective laws in the United States for protecting species is the Endangered Species Act of 1973. This law is often at the center of controversy between environmental interests and economic interests. As a result, compromises are often reached in which some habitat and species are protected in exchange for allowing some limited development.

8. International agreements and conventions protecting biological diversity are needed for the following reasons: species migrate across borders, there is an international trade in biological products, the benefits of biological diversity are of international importance, and the threats to diversity are often international in scope. The Convention on International Trade in Endangered Species (CITES) was enacted to regulate and monitor trade in endangered species. Many countries, particularly those in the European Union, use these international agreements to protect species within their own borders.

Suggested Readings

Akçakaya, H. R., M. A. Burgman and L. R. Ginzburg. 1999. *Applied Population Ecology: Principles and Computer Exercises Using RAMAS® EcoLab.* Sinauer Associates, Sunderland, MA. Quantitative principles of population biology are applied to conservation biology, using the *RAMAS® EcoLab* software.

Avise, J. C. and J. L. Hamrick (eds.). 1996. *Conservation Genetics: Case Histories from Nature.* Chapman and Hall, New York. Leading experts present current knowledge of genetics for many groups of organisms.

Clemmons, J. R. and R. Buchholz (eds.). 1997. *Behavioral Approaches to Conservation in the Wild.* Cambridge University Press, New York. Conservation projects need to pay careful attention to animal behavior and to adjust management practices accordingly.

Falk, D. A., C. I. Millar and M. Olwell (eds.). 1996. *Restoring Diversity: Strategies for Reintroduction of Endangered Plants.* Island Press, Washington D.C. Policy, biology, legal issues, and case studies.

Given, D. 1994. *Principles and Practices of Plant Conservation.* Timber Press, Portland, OR. Review of current botanical approaches.

IUCN. 1996. *1996 IUCN Red List of Threatened Animals.* IUCN, Gland, Switzerland. Evaluation of 5205 animal species using new quantitative criteria, plus a wealth of other information.

Norton, B. G., M. Hutchins, E. F. Stevens and T. L. Maple. 1995. *Ethics on the Ark: Zoos, Animal Welfare, and Wildlife Conservation.* Smithsonian Institution Press, Washington, D.C. Vigorous examination of the ethical issues confronting modern zoos.

Noss, R. F., M. A. O'Connell and D. D. Murphy. 1997. *The Science of Conservation Planning: Habitat Conservation under the Endangered Species Act.* Island Press, Washington, D.C. Scientific research and principles need to have a more important role in conservation planning.

Poole, J. 1996. *Coming of Age with Elephants: A Memoir.* Hyperion, New York. Personal account of how studies of elephants in Kenya led to involvement in their protection.

Primack, R. and B. Drayton. 1997. The experimental ecology of reintroduction. *Plant Talk* 11 (October): 25–28. Practical advice for plant reintroductions. Consult this outstanding and beautiful magazine for news on plant conservation.

Rhoades, R. E. 1991. World's food supply at risk. *National Geographic* 179 (April): 74–105. Beautifully illustrated popular account of the decline of traditional agricultural varieties and the need for seed banks.

Schaller, G. B. 1993. *The Last Panda.* University of Chicago Press, Chicago. Leading wildlife biologist describes a conservation program set within a complex political and economic landscape.

Stein, B. A. and S. R. Flack. 1997. *Species Report Card: The State of U.S. Plants and Animals.* The Nature Conservancy, Arlington, VA. The Nature Conservancy approach applied to the relatively well-known U.S. biota.

TRAFFIC USA. World Wildlife Fund, Washington, D.C. An informative newsletter covering the international trade in wildlife and wildlife products, with an emphasis on CITES activities.

Wilson, D. E., F. R. Cole, J. D. Nichols and R. Rudran. 1996. *Measuring and Monitoring Biological Diversity: Standard Methods for Mammals.* Biological Diversity Handbook Series. Smithsonian Institution Press, Washington, D.C. Technical manual for field biologists.

Young, O. R. (ed.). 1999. *The Effectiveness of International Environmental Regimes: Causal Connections and Behavioral Mechanisms.* MIT Press, Cambridge, MA. Analysis of why some agreements work and others do not.

Chapter 4

Conservation at the Community Level

Protecting habitats that contain healthy, intact biological communities is the most effective way to preserve overall biological diversity. One could even argue that it is ultimately the only way to preserve species, because we have the resources and knowledge to maintain only a small minority of the world's species in captivity. Four means of preserving biological communities are the establishment of protected areas, the effective management of those areas, implementation of conservation measures outside protected areas, and the restoration of biological communities in degraded habitats.

Biological communities vary from those few that are virtually unaffected by human activities—such as communities found on the ocean floor or in the most remote parts of the Amazon rain forest—to those that are heavily modified by human activity—such as agricultural land, cities, and artificial ponds. But even in the most remote areas of the world, human influence is apparent in the form of rising carbon dioxide levels, chemical pollution, and the collection of valuable natural products; conversely, even in the most modified of human environments there are often remnants of the original biota. Habitats with intermediate levels of disturbance present some of the most interesting challenges and opportunities for conservation biology because they often cover large areas. Considerable biological diversity may remain in selectively logged tropical forests, heavily fished oceans and seas, and grasslands grazed by domestic livestock (Western 1989; Redford 1992). When a conservation area is established, the right compromise must be found between protecting biological diversity and ecosystem functions and satisfying the immediate and long-term needs for resources of the local human community and the national government.

Protected Areas

One of the most critical steps in preserving biological communities is the establishment of legally designated protected areas. While legislation and purchases of land do not by themselves ensure habitat preservation, they represent an important starting point.

Protected areas can be established in a variety of ways, but the two most common mechanisms are government action (often at a national level, but also at regional and local levels) and land purchases by private individuals and conservation organizations. Governments can set aside land for protected areas and enact laws that allow varying degrees of commercial resource use, traditional use by local people, and recreational use in those areas. Many protected areas have also been established by private conservation organizations, such as The Nature Conservancy and the Audubon Society (The Nature Conservancy 1996). An increasingly common pattern is that of a partnership between the government of a developing country and international conservation organizations, multinational banks, and governments of developed countries; in such partnerships, the conservation organizations often provide funding, training, and scientific and management expertise to assist the developing country in establishing a new protected area. This type of collaborative effort is increasing due to new funding provided by the Global Environment Facility (GEF), which was created by the World Bank and agencies of the United Nations (see Chapter 5).

Protected areas are also established by traditional societies seeking to maintain their cultures. National governments have recognized traditional societies' rights to land in many countries, including the United States, Canada, Brazil, and Malaysia, though often not before conflict in the courts, in the press, and on the land under debate. In many cases, assertions of local rights to traditional lands have involved violent confrontations with the existing authorities seeking to develop the land; sometimes these resulted in the loss of life (Gadgil and Guha 1992; Western et al. 1994).

Once land comes under protection, decisions must be made regarding how much human disturbance will be allowed. The International Union for the Conservation of Nature has developed a system of classification for protected areas that covers a range from minimal to intensive use of the habitat by humans (IUCN 1994a):

I. **Strict nature reserves and wilderness areas** protect species and natural processes in an as undisturbed state as possible. These areas provide representative examples of biological diversity for scientific study, education, and environmental monitoring.

II. **National parks** are large areas of scenic and natural beauty maintained to provide protection for one or more ecosystems and for scientific, educational, and recreational use; they are not usually used for commercial extraction of resources.

III. **National monuments and landmarks** are smaller reserves designed to preserve unique biological, geological, or cultural features of special interest.

IV. **Managed wildlife sanctuaries and nature reserves** are similar to strict nature reserves, but some human manipulation—such as removing exotic species and setting managed fires—may be carried out to maintain the characteristics of the community. Some controlled harvesting may be permitted.

V. **Protected landscapes and seascapes** allow nondestructive traditional uses of the environment by resident people, particularly where these uses have produced an area of distinctive cultural, aesthetic, and ecological characteristics. Examples include fishing villages, orchards, and grazing land. Such places provide special opportunities for tourism and recreation.

VI. **Managed resource protected areas** allow for sustained production of natural resources, including water, wildlife, grazing for livestock, timber, tourism, and fishing, in a manner that ensures the preservation of biological diversity. These areas are often large and may include both modern and traditional uses of natural resources.

Of these categories, the first five can be considered truly **protected areas**, with the habitat managed primarily for biological diversity. A stricter definition would include only the first three categories. Areas in the last category are not managed primarily for biological diversity, though this may be a secondary management goal. **Managed areas** are particularly significant because they are often much larger than protected areas, they still may contain many or even most of their original species, and protected areas are often embedded in a matrix of managed areas.

Existing protected areas

As of 1998, around 4500 strictly protected areas (IUCN categories I–III) had been designated worldwide, covering some 500 million ha, with an additional 5899 partially protected areas (IUCN categories IV–VI) covering 348 million ha (Table 4.1; WRI 1998). Although this amount may seem impressive, it represents only about 6% of the Earth's total land surface. Only 4% of the Earth's total land surface is in the strictly protected categories of scientific reserves, national parks, and national monuments. In addition, the coverage of protected areas varies dramatically among countries, with high proportions of land protection in countries such as Germany (25%), Austria (25%), and the United Kingdom (19%), but surprisingly low proportions in others, including Russia (1.2%), Greece (0.8%), and Turkey (0.3%). The world's largest single protected area is in Greenland and covers 97 million ha. The figures for individual countries and continents are only approximations because sometimes the laws protecting national parks and wildlife sanctuaries are not actually enforced and sometimes sections of resource reserves and multiple-use management areas are carefully protected in practice. Examples of the latter are the sections within U.S. National Forests that are designated as wilderness areas.

Protected areas will never cover more than a small percentage of the Earth's total land surface—perhaps 7%–10% or slightly more—

TABLE 4.1 Protected and managed areas in the world's geographical regions[a]

| Region | Totally protected areas (IUCN categories I–III) | | Partially protected areas (IUCN categories IV–V) | | % of land area protected |
	Number of areas	Size (× 1000 ha)	Number of areas	Size (× 1000 ha)	
Africa	300	90,091	446	63,952	5.2
Asia	629	105,553	1104	57,324	5.3
North America	1243	113,370	1090	101,344	11.7
Central America	200	8346	214	6446	5.6
South America	487	81,080	323	47,933	7.4
Europe[b]	615	47,665	2538	57,544	4.7
Oceania[c]	1028	53,341	184	7,041	7.1
World	4502	499,446	5899	348,433	6.4

Source: WRI 1998.

[a]Includes only lands protected by national governments over 1000 ha in area; does not include private or locally protected sites; also does not include Antarctica or Greenland.

[b]Includes the Russian Federation.

[c]Includes Australia, New Zealand, Papua New Guinea, Fiji, and the Solomon Islands.

due to the demands of human societies for natural resources. The establishment of new protected areas peaked in the period from 1970 to 1975 and has been declining since then, probably because remaining land has already been designated for other purposes (McNeely et al. 1994). This limited area of protected habitat emphasizes the biological significance of the 10%–20% of the land that is managed for resource production. In the United States, for example, the Forest Service and the Bureau of Land Management together manage 24% of the land; in Costa Rica, about 17% of the land is managed as forest and Indian reserves.

Marine conservation efforts have lagged behind terrestrial conservation. Currently only 1% of the marine environment is included in protected areas, yet 20% of the marine environment needs to be protected in order to manage declining commercial fishing stocks (Costanza et al. 1998). Efforts to protect marine biological diversity have been hindered by the difficulty of identifying distinct biological communities and by the widespread migration and dispersal of marine species. In addition, opposition from fishing interests, the widespread impact of marine pollution, the difficulties of concluding international agreements, and the problems of policing large areas have also slowed efforts to establish effective marine reserves. However, the conservation community has identified marine conservation as a high priority, and urgent efforts are currently underway to protect marine biological diversity by establishing marine parks. Protecting the nursery grounds of commercial species and maintaining high-quality areas for recreational activities such as diving, swimming, and fishing, are among the main reasons for establishing these reserves. Over 1300 marine and coastal protected areas have been established worldwide, protecting about 800,000 km^2 (Agardy 1997). Accounting for half of the total are the three largest protected marine areas: the Great Barrier Reef Marine Park in Australia, the Galápagos Marine Park in Ecuador, and the Netherlands' North Sea Reserve.

The effectiveness of protected areas

If protected areas cover only a small percentage of the Earth, how effective will they be at preserving the world's species? Concentrations of species occur at particular places in the landscape: along elevational gradients, at places where different geological formations are juxtaposed, in areas that are geologically old, and at locations that have an abundance of critical natural resources (e.g., streams and water holes in otherwise dry habitats; caves that can be used by bats and birds for nesting) (Carroll 1992). Often a landscape contains large expanses of a fairly uniform habitat type and only a few small areas of rare habitat types. Protecting biological diversity in this case will probably depend not so much on preserving large areas of the common habitat type as on including representatives of all the habitat types in a system of protected areas. The following examples illustrate the potential effectiveness of protected areas of limited extent.

- The Indonesian government plans to protect populations of all native bird and primate species within its system of national parks and reserves. This goal will be accomplished by increasing the extent of protected areas from 3.5% to about 10% of Indonesia's land area.

- In most of the large tropical African countries, the majority of the native bird species have populations inside protected areas (Table 4.2). For example, the Democratic Republic of the Congo has over 1000 bird species, and 89% of them occur in the 3.9% of the land area under protection. Similarly, 85% of Kenya's birds are protected in the 5.4% of the land area included in parks (Sayer and Stuart 1988).

- A dramatic example of the importance of small protected areas is Santa Rosa Park in northwestern Costa Rica. This park covers only 0.2% of the area of Costa Rica, yet it contains breeding populations of 55% of the country's 135 species of sphingid moths. Santa Rosa Park is included within the new 82,500 ha Guanacaste National Park, which is predicted to contain populations of almost every sphingid moth (Janzen 1988).

TABLE 4.2 Percentage of bird species found within protected areas for selected African nations

Country	% of national land area protected	Number of bird species	% of bird species found in protected areas
Cameroon	3.6	848	76
Côte d'Ivoire	6.2	683	83
Democratic Republic of the Congo	3.9	1086	89
Ghana	5.1	721	77
Kenya	5.4	1064	85
Malawi	11.3	624	78
Nigeria	1.1	831	86
Somalia	0.5	639	47
Tanzania	12.0	1016	82
Uganda	6.7	989	89
Zambia	8.6	728	88
Zimbabwe	7.1	635	92

Source: From Sayer and Stuart 1988.

These examples clearly show that well-selected protected areas can include many, if not most, of the species in a country. However, the long-term future of many species in these reserves remains in doubt. Populations of many species may be so reduced in size that their eventual fate is extinction. Consequently, while the number of species existing in a relatively new park is important as an indicator of the park's potential, the real value of the park is its ability to support viable long-term populations of species. In this regard, the size of the park and the way it is managed are critical.

Establishing priorities for protection

In a crowded world with limited funding, priorities must be established for conserving biological diversity and, most importantly, individual species. While some conservationists would argue that no species should ever be lost, the reality is that species are being lost every day. The real question is how this loss of species can be minimized given the financial and human resources available. The interrelated questions that must be addressed by conservation planners are: *What* needs to be protected? *Where* should it be protected? and *How* should it be protected (Johnson 1995)? Three criteria can be used in setting conservation priorities for the protection of species and communities.

1. *Distinctiveness.* A biological community is given higher priority for conservation if it is composed primarily of rare endemic species than if it is composed primarily of common, widespread species. A species is often given more conservation value if it is taxonomically unique—that is, the only species in its genus or family—than if it is a member of a genus with many species (Vane-Wright et al. 1994).

2. *Endangerment.* Species in danger of extinction are of greater concern than species that are not threatened with extinction; thus the whooping crane, with only about 155 individuals, is of greater concern than the sandhill crane, with approximately 500,000 individuals. Biological communities threatened with imminent destruction are also given priority.

3. *Utility.* Species that have present or potential value to people are given more conservation value than species of no obvious use to people. For example, wild relatives of wheat, which are potentially useful in developing improved cultivated varieties, are given a higher priority than species of grass that are not known to be related to any economically important plant. Biological communities of major economic value, such as coastal wetlands, may have greater priority for protection than less valuable communities, such as dry scrubland.

The Komodo dragon of Indonesia (Figure 4.1) is an example of a species that would be a conservation priority using all three criteria: It is the world's largest lizard (distinctive); it occurs on only a few small islands of a rapidly developing nation (endangered); and it has major potential as a tourist attraction, as well as being of great scientific interest (utility). Using these criteria, several priority systems have been developed at both national and international scales to target both species and communities. These approaches are generally complementary; they differ more in emphasis than in fundamental principles. In any case, priorities for new protected areas need to be established so that personnel and financial resources are directed to the most critical problems.

Species approaches. Protected areas can be established to conserve individual species. Many national parks have been created to protect "charismatic megafauna" or special **focal species**, such as tigers and macaws, which capture public attention, have symbolic value, and are crucial to ecotourism. Other protected areas have been established for endangered **indicator species,** whose presence demonstrates the health of the ecosystem; protecting these species also preserves whole communities that may consist of thousands of other species along with the underlying ecosystem processes (Paoletti 1999; Schwartz 1999). For example, Project Tiger in India was

4.1 The carnivorous Komodo dragon (*Varanus komodoensis*) of Indonesia is the largest living monitor lizard. Many people feel it possesses unique status, and that protecting this endangered species is a conservation priority. (Photograph by Jessie Cohen, National Zoological Park, Smithsonian Institution.)

started in 1973 after a census revealed that the Indian tiger was in imminent danger of extinction. Project Tiger has helped to provide attention, funding, and a management philosophy for national parks in India. The establishment of 18 Project Tiger reserves, combined with strict protection measures, has slowed the rapid decline in the number of tigers and the degradation of the biological communities in which they live (Ward 1992).

Identifying geographical areas of high conservation priority is the first step in developing survival plans for individual species. In the Americas, the Natural Heritage Programs and Conservation Data Centers, which are associated with government agencies, are acquiring data on the past and present distribution and ecology of rare and endangered species from all 50 U.S. states, 3 Canadian provinces, and 13 Latin American countries (Jenkins 1996). This information is being used to target new localities for conservation. Another important program is the IUCN Species Survival Commission Action Plans. Approximately 2000 scientists have been organized into 80 specialist groups to provide evaluations and recommendations for mammals, birds, invertebrates, reptiles, fishes, and plants (Species Survival Commission 1990). One group, for example, produced an Action Plan for Asian primates, in which priority rankings were developed for 64 species based on degree of threat, taxonomic uniqueness, and association with other threatened primates. The new protected areas and the specific actions needed to protect these primates were highlighted for the benefit of policy makers and conservation organizations.

Community and ecosystem approaches. A number of conservationists have argued that communities and ecosystems, rather than species, should be the target of conservation efforts (Reid 1992; Grumbine 1994a). Conservation of communities can preserve large numbers of species in a self-maintaining unit, whereas targeted species rescues are often difficult, expensive, and ineffective. Spending $1 million on habitat protection and management might preserve more species in the long run than spending the same amount on an intensive effort to save just one conspicuous species. Also, the ecosystem value of habitat preservation often provides a strong economic argument for action.

New protected areas are often placed to ensure that **representative sites** of as many types of biological communities as possible are protected. A representative site includes the species and environmental conditions characteristic of the biological community. Determining which countries and areas of the world and which biological communities have adequate conservation protection and which urgently need additional protection is critical to the world conservation movement. Resources, research, and publicity must be directed to areas of the world that require additional protection. Regions throughout the world are being evaluated for the current percentage of areas under protection, threats, need for action, and conservation importance (McNeely et al. 1994; Ricketts et al. 1999). Analysis so far indicates that protecting lake systems and temperate grasslands are among the highest conservation priorities.

Gap analysis. One way to determine the effectiveness of ecosystem and community conservation programs is to compare biodiversity priorities with existing and proposed protected areas (Scott and Csuti 1996; Olson and Dinerstein 1998). This comparison can identify "gaps" in biodiversity preservation that need to be filled with new protected areas. On an international scale, **gap analysis** is helping to identify for protection representative examples of all of the world's seven terrestrial biogeographic regions and 193 biological provinces (regions with concentrations of endemic species). Although all major biogeographic regions of the world have some protected areas (see Table 4.1), 10 of the 193 biological provinces have no protected areas, and 38 have less than 1% of their area under protection (McNeely et al. 1994).

At the national level, biological diversity is protected most efficiently by ensuring that all major ecosystem types are included in a system of protected areas. In the United States, various federal and state agencies, often led by personnel from the Natural Heritage Programs, are involved in an intensive "bottom-up" effort to survey and classify ecosystems on a local level as part of a program to protect biological diversity. An alternative, "top-down" approach is to compare a detailed vegetation map with a map of lands under government protection (Crumpacker et al. 1988). Such an analysis has shown that 11 distinctive biological communities are represented on government lands only by small areas either because they are naturally rare or because they have been largely destroyed. Most of the unrepresented ecosystems are in central or southern Texas (e.g., mesquite savannah) and in Hawaii (e.g., mixed guava forest). These community types are being highlighted in conservation efforts and included if possible in new protected areas.

Geographic Information Systems (GIS) represent the latest development in gap analysis technology, using computers to integrate the wealth of spatial data on the natural environment with information on species distributions (Wright et al. 1994; Kremen et al. 1999). GIS analyses make it possible to highlight critical areas that need to be included within national parks and areas that should be avoided by development projects. The basic GIS approach involves storing, displaying, and manipulating many types of mapped data, such as vegetation types, climate, soils, topography, geology, hydrology, species distributions, human settlements, and resource use

(Figure 4.2). This approach can point up correlations among the abiotic, biotic, and socioeconomic elements of the landscape, help plan parks that include the greatest diversity of biological communities, and even suggest potential sites in which to search for and protect rare species. Aerial photographs and satellite imagery are additional sources of data for GIS analysis. In particular, a series of images taken over time can reveal patterns of habitat destruction that need prompt attention.

Centers of biodiversity. In order to establish priorities for conservation efforts, the World Conservation Monitoring Centre, Birdlife International, Conservation International, and others have attempted to identify key areas of the world that have great biological diversity and high levels of endemism and are under immediate threat of species extinctions and habitat destruction: so-called "hot spots" for preservation (Figure 4.3; Table 4.3). Using these criteria, Mittermeier et al. (1999) identified 25 global hot spots that together encompass the entire ranges of 44% of the world's plant species, 28% of the bird species, 30% of the mammal species, 38% of the rep-

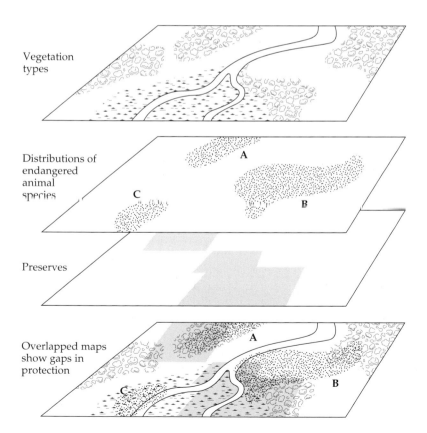

Vegetation types

Distributions of endangered animal species

Preserves

Overlapped maps show gaps in protection

A

B

C

A

B

C

4.2 Geographic Information Systems (GIS) provide a method for integrating a wide variety of data for analysis and display on maps. In this example, vegetation types, animal distributions, and preserved areas are overlapped to highlight areas that need additional protection. The distribution of Species A is predominantly in a reserve, Species B is only protected to a limited extent, and Species C is found entirely outside of the reserves. (After Scott et al. 1991.)

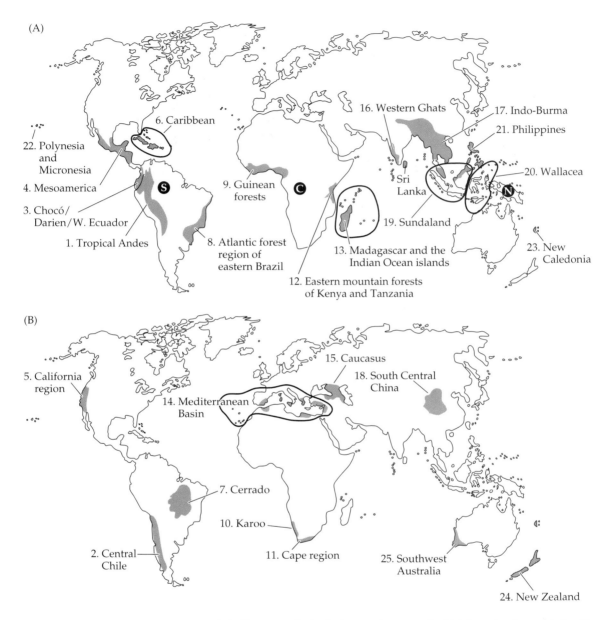

4.3 (A) Fifteen tropical rain forest "hot spots" of high endemism and significant threat of imminent extinctions. Numbered regions correspond to Table 4.3. The circled areas enclose four island hot spots: Caribbean, Madagascar, and Indian Ocean islands, plus Sundaland and Wallacea regions. The Polynesia/Micronesia region covers a large number of Pacific Ocean islands, including the Hawaiian Islands, Fiji, Samoa, French Polynesia, and the Marianas. Circled letters indicate the only three remaining tropical forest wilderness areas of any extent: S = South America, C = Congo Basin, N = New Guinea. (B) Ten "hot spots" in other ecosystems. The circled area encloses the Mediterranean Basin. (After Mittermeier et al. 1999.)

TABLE 4.3   A comparison of twenty-five global hot spots

	Original extent (× 1000 km²)	Percentage remaining	Percentage protected	Number of		
				Plants	Birds	Mammals
1. Tropical Andes	1258	25.0	6.3	45,000	1666	414
2. Central Chile	300	30.0	3.1	3429	198	56
3. Chocó/Darien/ Western Ecuador	261	24.2	6.3	9,000	830	235
4. Mesoamerica	1155	20.0	12.0	24,000	1193	521
5. California region	324	24.7	9.7	4426	341	145
6. Caribbean	264	11.3	15.6	12,000	668	164
7. Brazilian Cerrado	1783	20.0	1.2	10,000	837	161
8. Atlantic forest of Brazil	1227	7.5	2.7	20,000	620	261
9. Guinean forests of West Africa	1265	10.0	1.6	9000	514	551
10. South African karoo	112	27.0	2.1	4849	269	78
11. Cape region of South Africa	74	24.3	19.0	8200	288	127
12. Eastern mountain forests of Kenya and Tanzania	30	6.7	16.9	4000	585	183
13. Madagascar and Indian Ocean islands	594	9.9	1.9	12,000	359	112
14. Mediterranean Basin	2362	4.7	1.8	25,000	345	184
15. Caucasus region east of the Black Sea	500	10.0	2.8	6,300	389	152
16. Western Ghats and Sri Lanka	182	6.8	10.4	4780	528	140
17. Indo-Burma	2060	4.9	7.8	13,500	1170	329
18. Mountains of south central China	800	8.0	2.1	12,000	686	300
19. Sundaland Island region	1600	7.8	5.6	25,000	815	328
20. Wallacea Island region	347	15.0	5.9	10,000	697	201
21. Philippines	301	8.0	1.3	7620	556	201
22. Polynesia/Micronesia	46	21.8	10.7	6557	254	16
23. New Caledonia	19	28.0	2.8	3332	116	9
24. New Zealand	271	22.0	19.2	2300	149	3
25. Southwest Australia	310	10.8	10.8	5469	181	54

Source: From Mittermeier et al. 1999.

tile species, and 54% of the amphibian species, on only 1.4% of the Earth's total land surface. Because these hot spots also include more widespread species, they actually include about two-thirds of all non-fish vertebrates on the planet. Many of these hot spots are tropical rain forest areas, such as the Atlantic Coast of Brazil, the Chocó/Darien/Western Ecuador region, Mesoamerica, the Guinean forests of West Africa, the Western Ghats of India, and the Indo-Burma region. Island areas are also among these hot spots, including the Caribbean region, Madagascar, Sri Lanka, the Sundaland and Wallacea regions of Malaysia and Indonesia, the Philippines,

New Caledonia, New Zealand, and the Polynesia region. Hot spots are also located in warm, seasonally dry areas in the temperate zone, such as the Mediterranean Basin, the California region, central Chile, the Cape region of South Africa, the Caucasus region, and southwest Australia. Remaining areas are the dry forests and savannahs of the Brazilian Cerrado, the eastern mountains of Kenya and Tanzania, the tropical Andes, and the mountains of south central China. These habitats originally covered 17 million km^2 but are now intact on only 2 million km^2, and protected on only 888,789 km^2, only 0.60% of the Earth's total surface. One of the Earth's major centers of biodiversity is the tropical Andes, in which 45,000 plant species, 1666 bird species, 414 mammal species, 479 reptile species, and 830 amphibian species persist in tropical forests and high-altitude grasslands on less than one quarter of one percent of the Earth's total land surface. The hot spot approach can also be applied to individual countries. In the United States, hot spots for endangered species occur in the Hawaiian Islands, the southern Appalachians, the arid Southwest, and the coastal areas of the lower 48 states (Flather et al. 1998; Dobson et al. 1997).

Another valuable approach has been to identify seventeen "megadiversity" countries (out of a global total of more than 230 countries) that together contain 60%–70% of the world's biological diversity: Mexico, Colombia, Brazil, Peru, Ecuador, Venezuela, the United States, the Democratic Republic of the Congo, South Africa, Madagascar, Indonesia, Malaysia, Philippines, India, China, Papua New Guinea, and Australia. These countries are possible targets for increased conservation attention and funding (Table 4.4; Mittermeier et al. 1997).

Certain organisms can be used as **biological diversity indicators** when specific data about whole communities are unavailable. Diversity in plants and birds, for example, are sometimes, but not always, good indicators of the diversity of a community (Ricketts et al. 1999). Several analyses have put this principle into practice. The IUCN Plant Conservation Office in England is identifying and documenting about 250 global centers of plant diversity with large concentrations of species (WWF/IUCN 1997). The International Council for Bird Protection (ICBP) is identifying localities with large concentrations of birds that have restricted ranges, called Endemic Bird Areas (EBAs) (Stattersfield et al. 1998). To date 218 such localities, containing 2451 restricted-range bird species, have been identified. Many of these localities are islands and isolated mountain ranges that also have many endemic species of lizards, butterflies, and trees, and thus represent priorities for conservation. Further analysis has highlighted EBAs that contain no protected areas and require urgent conservation measures. Another

TABLE 4.4 "Top ten" rankings for countries with the largest number of species from well-known groups

Rank	Higher plants[a]	Mammals	Birds	Reptiles	Amphibians	Freshwater fish	Butterflies
1	Brazil 53,000	Brazil 524	Colombia 1815	Australia 755	Colombia 583	Brazil >3000	Peru 3532
2	Colombia 47,000	Indonesia 515	Peru 1703	Mexico 717	Brazil 517	Colombia >1500	Brazil 3,132
3	Indonesia 37,000	China 499	Brazil 1622	Colombia 520	Ecuador 402	Indonesia 1400	Colombia 3100
4	China 28,000	Colombia 456	Ecuador 1559	Indonesia 511	Mexico 284	Venezuela 1250	Bolivia 3000
5	Mexico 24,000	Mexico 450	Indonesia 1531	Brazil 468	China 274	China 1010	Venezuela 2316
6	South Africa 23,000	USA 428	Venezuela 1360	India 408	Indonesia 270	DRC 962	Mexico 2237
7	Ecuador 19,000	DRC 415	India 1258	China 387	Peru 241	Peru 855	Ecuador 2200
8	Peru 19,000	India 350	Bolivia 1257	Ecuador 374	India 206	Tanzania 800	Indonesia 1900
9	PNG 18,000	Peru 344	China 1244	PNG 305	Venezuela 204	USA 790	DRC 1650
10	Venezuela 18,000	Uganda 315	DRC 1094	Madagascar 300	PNG 200	India 750	Cameroon 1550

Source: From Mittermeier et al. 1997.
Note: PNG = Papua New Guinea; DRC = Democratic Republic of the Congo; USA = United States of America.
[a]Flowering plants, gymnosperms, and ferns.

approach involves using well-known indicator groups, such as birds, mammals, plants, and butterflies, to protect **complementary areas**, which are selected for containing the greatest overall number of species (Howard et al. 1998; Balmford and Gaston 1999). In this approach, new conservation areas are added not based solely on their own characteristics, but according to how well they add to the total number of species and biological communities already under protection.

Wilderness areas. Large blocks of land that have been minimally affected by human activity, have a low human population density, and are not likely to be developed in the near future, are perhaps the only places on Earth where large mammals can survive in the wild. These wilderness areas could potentially remain as controls to show what natural communities are like with minimal human influence. In the United States, proponents of the Wildlands Project are advocating the management of whole ecosystems to preserve viable populations of large carnivores, such as grizzly bears, wolves, and large cats (Noss and Cooperrider 1994). Three tropical wilderness areas also have been identified and established as conservation priorities (see Figure 4.3A; Bryant et al. 1997).

- *South America:* One arc of wilderness, containing rain forest, savannah, and mountains—but few people—runs through the southern Guianas, southern Venezuela, northern Brazil, Colombia, Ecuador, Peru, and Bolivia.

- *Africa:* A large area of equatorial Africa, centered on the Congo basin, has a low population density and undisturbed habitat. This area includes large portions of Gabon, the Republic of the Congo, and the Democratic Republic of the Congo. Warfare and a lack of government control prevents effective conservation activities in large parts of the region.

- *New Guinea:* The island of New Guinea has the largest tracts of undisturbed forest in the Asian Pacific region, despite the effects of logging, mining, and transmigration programs. The eastern half of the island is the independent nation of Papua New Guinea, with 3.9 million people in 462,840 km^2. The western half of the island, Irian Jaya, is a state of Indonesia and has a population of only 1.4 million people in 345,670 km^2. Large tracts of forest also occur on the island of Borneo, but logging, plantation agriculture, fire, an expanding human population, and the development of a transportation network are rapidly reducing the area of undisturbed forest there.

International agreements

Habitat conventions at the international level complement species conventions, such as CITES, by emphasizing unique ecosystem features that need to be protected. Within these habitats, multitudes of individual species can be protected. Three of the most important such conventions are the Convention on Wetlands of International Importance Especially as Waterfowl Habitat (also known as the Ramsar Convention on Wetlands), the Convention Concerning the Protection of the World Cultural and Natural Heritage (or the World Heritage Convention), and the UNESCO Man and the Biosphere Program (or the Biosphere Reserves Program). Countries designating protected areas under these conventions voluntarily agree to administer them under the terms of the conventions; countries do not agree to give up sovereignty over the areas to an international body but retain full control over them.

The Ramsar Convention on Wetlands was established in 1971 to halt the continued destruction of wetlands, particularly those that support migratory waterfowl, and to recognize the ecological, scientific, economic, cultural, and recreational values of wetlands (Hails 1996). The Ramsar Convention covers freshwater, estuarine, and coastal marine habitats, and includes more than 895 sites with a total area of over 66 million ha. Each of the 94 signing countries agreed to conserve and protect their wetland resources and to designate for conservation purposes at least one wetland site of international significance.

The Convention Concerning the Protection of the World Cultural and Natural Heritage is associated with UNESCO, IUCN, and the International Council on Monuments and Sites (von Droste et al. 1995). This convention has received unusually wide support, with 109 countries participating, among the most of any conservation convention. The goal of the convention is to protect natural areas of international significance through its World Heritage Site Program. The convention is unusual because it emphasizes the cultural as well as the biological significance of natural areas and recognizes that the world community has an obligation to support the sites financially. Included in the list of 126 World Heritage Sites, covering 127 million ha, are some of the world's premier conservation areas: Serengeti National Park (Tanzania), Sinharaja Forest Reserve (Sri Lanka), Iguaçú National Park (Brazil), Manu National Park (Peru), Queensland Rain Forest (Australia), Great Smokies National Park (U.S.), and Komodo National Park (Indonesia) (Figure 4.4).

(A)

(B)

4.4 Many of the world's most revered and well-known conservation areas are protected as national parks. (A) Iguaçú Falls, Iguaçú National Park, Brazil. (B) Starfish thrive in clear water off Kanawa Island, one of the islands of Komodo National Park, Indonesia. Both of the parks have been designated as World Heritage Sites. (Photographs © Josh Schacter.)

UNESCO's Man and the Biosphere Program (MAB) created an international network of Biosphere Reserves beginning in 1971. Biosphere Reserves are designed to be models demonstrating the compatibility of conservation efforts with sustainable development for the benefit of local people, as described in Chapter 5. As of 1998, a total of 337 reserves had been created in more than 80 countries, covering about 220 million ha, and including 47 reserves in the United States (Figure 4.5).

These three conventions have established a general consensus regarding appropriate conservation of habitat types. Major efforts are being made to organize these protected areas into international networks that can be used to address larger ecosystem and biodiversity questions at regional and global levels. More limited agreements protect unique ecosystems and habitats in particular regions, including the Western Hemisphere, the Antarctic flora and fauna, the South Pacific, Africa, and European wildlife and natural habitat (WRI 1994). Other international agreements have been signed to prevent or limit pollution that poses regional and international threats to the environment. The Convention on Long-Range Trans-Boundary Air Pollution in the European Region recognizes the role that long-range transport of air pollution plays in acid rain, lake acidification, and forest dieback. The Convention on the Protection of the Ozone Layer, signed in 1985, has been responsible for phasing out the use of chlorofluorocarbons, which have been linked to

4.5 Locations of recognized Biosphere Reserves (dots). A lack of reserves is apparent in such biologically important regions as New Guinea, the Indian subcontinent, South Africa, and Amazonia. (Data from UNESCO 1996.)

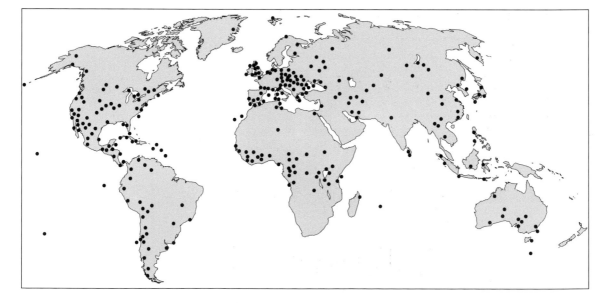

the destruction of the ozone layer and a resulting increase in the levels of harmful ultraviolet light. The United Nations Framework Convention on Climate Change of 1992 is designed to stabilize and eventually reduce emissions of greenhouse gases, principally carbon dioxide. The convention has been endorsed by more than 160 countries, and is reviewed at yearly conferences.

Marine pollution is another key area of concern because of the extensive areas of international waters not under national control and the ease with which pollutants released in one area can spread to another area (Norse 1993). Agreements covering marine pollution include the Convention on the Prevention of Marine Pollution by Dumping of Wastes and Other Matters, and the Regional Seas Conventions of the United Nations Environmental Programme (UNEP). Regional agreements cover the northeastern Atlantic, the Baltic, the Barents Sea, and other specific locations, particularly in the North Atlantic region.

Designing Protected Areas

The size and placement of protected areas throughout the world are often determined by the distribution of people, potential land values, the political efforts of conservation-minded citizens, and historical factors. In many cases, land is set aside for conservation protection because it has no immediate commercial value; protected areas are often sited on remote, infertile, resource-poor, uninhabited lands— that is, on "the lands that nobody wanted" (Pressey 1994; Wallis de Vries 1995). Small conservation areas are common in some large metropolitan areas, where they have been purchased by local governments or conservation organizations, or donated by wealthy citizens.

Although most parks and conservation areas have been acquired and created haphazardly, depending on the availability of money and land, a considerable body of ecological literature has been developed that provides guidelines for the most efficient ways to design conservation areas to protect biological diversity (Figure 4.6; Pressey et al. 1993; Shafer 1990, 1997). Some of these guidelines draw principles from the island biogeographical model of MacArthur and Wilson (1967; see Chapter 2). Conservation biologists use these guidelines in an attempt to address the following key questions of reserve design:

1. How large must nature reserves be to protect species?
2. Is it better to create a single large reserve or many smaller reserves?
3. How many individuals of an endangered species must be protected in a reserve to prevent extinction?

4.6 Principles of reserve design that have been proposed based on theories of island biogeography. Imagine that the reserves are "islands" of the original biological community surrounded by land that has been made uninhabitable by human activities such as farming, ranching, or industrial development. The practical application of these principles is still under study and debate, but in general the designs shown on the right are considered preferable to those shown on the left. (After Shafer 1997.)

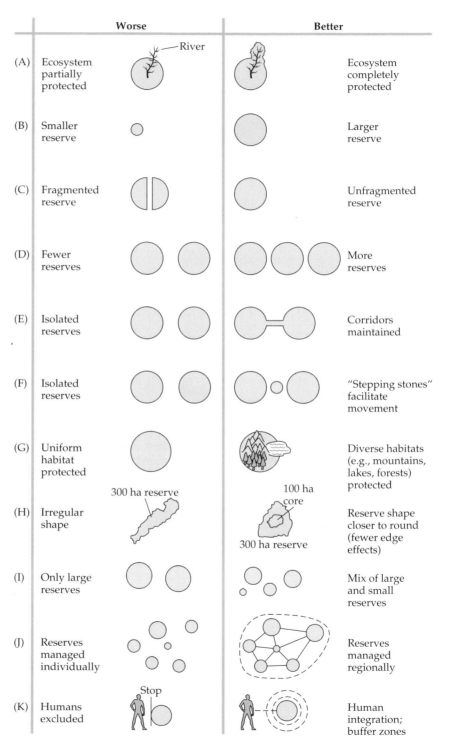

Worse		Better
(A) Ecosystem partially protected		Ecosystem completely protected
(B) Smaller reserve		Larger reserve
(C) Fragmented reserve		Unfragmented reserve
(D) Fewer reserves		More reserves
(E) Isolated reserves		Corridors maintained
(F) Isolated reserves		"Stepping stones" facilitate movement
(G) Uniform habitat protected		Diverse habitats (e.g., mountains, lakes, forests) protected
(H) Irregular shape		Reserve shape closer to round (fewer edge effects)
(I) Only large reserves		Mix of large and small reserves
(J) Reserves managed individually		Reserves managed regionally
(K) Humans excluded		Human integration; buffer zones

River

300 ha reserve

100 ha core

300 ha reserve

Stop

4. What is the best shape for a nature reserve?

5. When several reserves are created, should they be close together or far apart, and should they be isolated from one another or connected by corridors?

The guidelines of reserve design have proved to be of great interest to governments, corporations, and private landowners, who are being urged and mandated to manage their properties for both the commercial production of natural resources and for the protection of biological diversity. However, such guidelines do not necessarily represent universal solutions: Conservation biologists have been cautioned against providing simplistic, overly general guidelines for designing nature reserves, because every conservation situation requires special consideration (Ehrenfeld 1989). In addition, all would benefit from more communication between the academic scientists who are developing theories of nature reserve design and the managers who are actually creating new nature reserves (Prendergast et al. 1999). That said, guidelines for reserve design can provide a useful framework for constructing the best possible conservation area.

Reserve size

An early debate within conservation biology occurred over whether species richness is maximized in one large nature reserve or in several smaller ones of an equal total area (Diamond 1975; Simberloff and Abele 1982; Soulé and Simberloff 1986); this is known in the literature as the "SLOSS debate" (single large or several small). For example, is it better to set aside one reserve of 10,000 ha, or four reserves of 2500 ha each? The proponents of large reserves argue that only large reserves can contain sufficient numbers of large, wide-ranging, low-density species (such as large carnivores) to maintain long-term populations (Figure 4.7). Also, a large reserve minimizes edge effects, encompasses more species, and has greater habitat diversity than a small reserve. These advantages follow from island biogeographical theory and have been demonstrated in numerous surveys of animals and plants in parks. This viewpoint has three practical implications. First, when a new park is being established, it should be made as large as possible in order to preserve as many species as possible. Second, when possible, additional land adjacent to nature reserves should be acquired in order to increase the area of existing parks. And finally, if there is a choice of creating a new small park or a new large park in similar habitat types, the large park should be created. On the other hand, once a park is larger than a certain size, the number of new species added with each increase in area starts to decline. In such a situation, creat-

4.7 Population studies show that large parks and protected areas in Africa contain larger populations of each species than small parks; only the largest parks may contain long-term, viable populations of many vertebrate species. Each symbol represents an animal population. If the viable population size of a species is 1000 individuals (10^3; dashed line), parks of more than 100 (10^2) ha will be needed to protect small herbivores (e.g., rabbits, squirrels); parks of more than 10,000 (10^4) ha will be needed to protect large herbivores (e.g., zebras, giraffes); and parks of at least a million (10^6) ha will be needed to protect large carnivores (e.g., lions, hyenas). (After Schonewald-Cox 1983.)

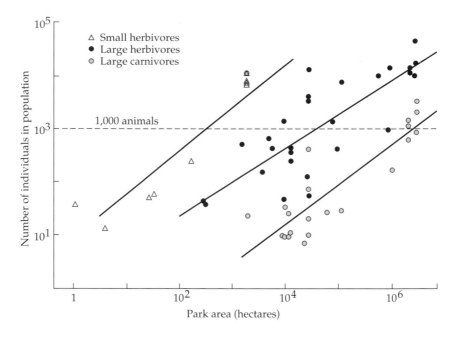

ing a second large park some distance away may be a better strategy for preserving additional species than adding to the existing park.

The extreme proponents of large reserves argue that small reserves should not be maintained because their inability to support long-term populations gives them little value for conservation purposes. Other conservation biologists argue that well-placed small reserves are able to include a greater variety of habitat types and more populations of rare species than would one large block of the same area (Simberloff and Gotelli 1984; Shafer 1995). Also, creating more reserves, even if they are small ones, prevents the possibility of a single catastrophic force, such as an exotic animal, a disease, or fire, destroying an entire population located in a single large reserve. In addition, small reserves located near populated areas can make excellent conservation education and nature study centers, furthering the long-range goals of conservation biology by developing a public awareness of important issues.

The consensus now seems to be that strategies on reserve size depend on the group of species under consideration, the land available, and the particular circumstances. It is accepted that large reserves are better able than small reserves to maintain many species because of the larger population sizes and greater variety of habitats they contain. However, well-managed small reserves also have value, particularly for the protection of many species of plants, invertebrates, and small vertebrates (Lesica and Allendorf 1992; Schwartz 1999). Often there is no choice other than to accept the challenge of managing species in small reserves because no additional land around them is available for conservation purposes. This is particularly true in places that have been intensively cultivated and settled for centuries, such as Europe, China, and Java. For example, Sweden has 1200 small nature reserves that average about 350 ha each, and small reserves account for 30%–40% of the protected area of the Netherlands (McNeely et al. 1994). Bukit Timah in Singapore is an example of a small, isolated nature reserve in an urban area that provides protection to numerous species. This 50 ha forest reserve represents 0.2% of Singapore's original forest area and has been isolated since 1860, yet it still protects 74% of the original flora, 72% of the bird species, and 56% of the fish (Corlett and Turner 1996).

Minimizing edge and fragmentation effects

It is generally agreed that parks should be designed to minimize harmful edge effects. Conservation areas that are rounded in shape minimize the edge-to-area ratio, and the center of such a park is farther from the edge than in other park shapes. Long, linear parks have the most edge, and all points in the park are close to the edge. Using these same arguments for parks with four straight sides, a square park is a better design than an elongated rectangle of the same area. These ideas have rarely, if ever, been implemented. Most parks have irregular shapes because land acquisition is typically a matter of opportunity rather than the completion of a geometric pattern.

Internal fragmentation of reserves by roads, fences, farming, logging, and other human activities should be avoided as much as possible because of the many negative effects fragmentation can have on species and populations (see Chapter 2). The forces promoting fragmentation are powerful, because protected areas are often the only undeveloped land available for new projects such as agriculture, dams, and residential areas. Government planners often locate transportation networks and other infrastructure in protected areas because they encounter less political opposition than if they choose to locate the projects on privately owned land. In the eastern United States, many parks near cities are crisscrossed by roads, railroad tracks, and power lines, which divide large areas of habitat into wedges, like pieces of a roughly cut pie.

Strategies exist for aggregating small nature reserves and other protected areas into larger conservation blocks. Nature reserves are often embedded in a larger matrix of habitat managed for resource extraction, such as timber forest, grazing land, or farmland. If the

4.8 Often national parks (shaded areas) are adjacent to other public lands (unshaded) that are managed by different agencies. The United States government is now managing large blocks of land that include national parks, national forests, and other federal lands as networks of natural areas in order to maintain populations of large and scarce wildlife. Four such networks are shown here. Privately owned land is shown in black. (After Salwasser et al. 1987.)

protection of biological diversity can be included as a secondary priority in the management of the production areas, then larger areas can be included in conservation management plans and the effects of fragmentation can be reduced (Figure 4.8). Also, by managing at a large scale, a greater representation of species and habitats can be protected. Whenever possible, populations of rare species should be managed as a large metapopulation to facilitate gene flow and migration among the populations (Soulé and Terborgh 1999). Cooperation between public and private landowners is particularly important in developed metropolitan areas, where there are often numerous small, isolated parks under the control of a variety of different government agencies and private organizations (Kohm and Franklin 1997). One such example is the Chicago Wilderness Project, in which 34 organizations, including local and national governments, private conservation organizations, citizen groups, zoos, and museums, are cooperating to preserve more than 56,000 ha (140,000 acres) of tallgrass prairies, wetlands, and woodlands in metropolitan Chicago (Yaffee et al. 1996).

Whenever possible, protected areas should include an entire ecosystem (such as a watershed, a lake, or a mountain range), because the ecosystem is the most appropriate unit of management.

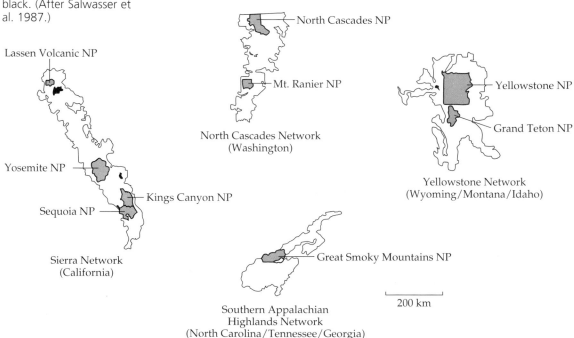

North Cascades NP

Mt. Ranier NP

North Cascades Network
(Washington)

Lassen Volcanic NP

Yosemite NP

Kings Canyon NP

Sequoia NP

Sierra Network
(California)

Yellowstone NP

Grand Teton NP

Yellowstone Network
(Wyoming/Montana/Idaho)

Great Smoky Mountains NP

200 km

Southern Appalachian
Highlands Network
(North Carolina/Tennessee/Georgia)

Damage to an unprotected portion of the ecosystem could threaten the health of the whole. Controlling the whole ecosystem allows park managers to defend it more effectively against destructive outside influences (Peres and Terborgh 1995).

Habitat corridors

One intriguing approach for managing a system of nature reserves has been to link isolated protected areas into one large system through the use of **habitat corridors**: strips of protected land running between the reserves (Simberloff et al. 1992; Rosenberg et al. 1997). Such habitat corridors, also known as conservation corridors or movement corridors, can allow plants and animals to disperse from one reserve to another, facilitating gene flow and colonization of suitable sites (Beier and Noss 1998). Corridors might also help to preserve animals that must migrate seasonally among a series of different habitats to obtain food; if these animals were confined to a single reserve, they could starve. This principle was put into practice in Costa Rica to link two wildlife reserves, the Braulio Carillo National Park and La Selva Biological Station. A 7700 ha corridor of forest several kilometers wide and 18 km long, known as La Zona Protectora, was set aside to provide an elevational link that allows at least 75 species of birds to migrate between the two important conservation areas (Bennett 1999).

Corridors are most obviously needed along known migration routes. For example, the Kibale Forest Game Corridor was established in 1926 to protect the migration route of game animals that stretches between Kibale Forest and Queen Elizabeth National Park in Uganda. In some cases, leaving small clumps of original habitat between large conservation areas may also serve to facilitate movement by creating a "stepping-stone" pattern. This is particularly important for migratory birds, which need to rest and feed. Where corridors already exist, they should be preserved. Many of the corridors that currently exist are along watercourses and may be biologically important habitats themselves. Specially constructed tunnels, culverts, bridges, and roads have been constructed to allow dispersal between habitats for mammals, lizards, and amphibians (Figure 4.9). On a larger scale, the Wildlands Project has proposed a plan to link all large protected areas in the United States by habitat corridors, creating a system that would allow large and currently declining mammals to coexist with human society.

Although corridors represent an appealing conservation strategy, they have some potential drawbacks. Corridors could facilitate the movement of pest species and disease, so that a single infestation could quickly spread to all of the connected nature reserves and cause the extinction of all populations of a rare species. Also, animals dispersing along corridors might be exposed to greater risks of

(A)

(B)

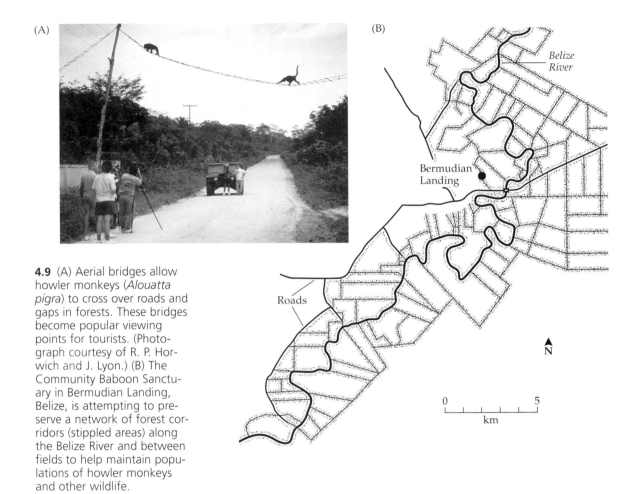

4.9 (A) Aerial bridges allow howler monkeys (*Alouatta pigra*) to cross over roads and gaps in forests. These bridges become popular viewing points for tourists. (Photograph courtesy of R. P. Horwich and J. Lyon.) (B) The Community Baboon Sanctuary in Bermudian Landing, Belize, is attempting to preserve a network of forest corridors (stippled areas) along the Belize River and between fields to help maintain populations of howler monkeys and other wildlife.

predation because human hunters, as well as animal predators, tend to concentrate on routes used by wildlife. Despite these concerns, the evidence to date tends to support the value of conservation corridors, though each case needs to be considered individually.

All of these theories of reserve design have been developed mainly with land vertebrates, higher plants, and large invertebrates in mind. The applicability of these ideas to aquatic nature reserves, where dispersal mechanisms are largely unknown, requires further investigation. Protection of marine parks requires particular attention to pollution control because of its subtle and widespread negative effects. Various countries in the Caribbean and the Pacific regions are actively developing marine protected areas. The entire Caribbean island of Bonaire is a protected marine park, with ecotourism emerging as a leading industry.

Landscape ecology and park design

The interaction of actual land use patterns, conservation theory, and reserve design is evident in the discipline of **landscape ecology**, which investigates patterns of habitat types on a regional scale and their influence on species distribution and ecosystem processes (Hansson et al. 1995; Forman 1995). A landscape is defined by Forman and Godron (1986) as an "area where a cluster of interacting stands or ecosystems is repeated in similar form" (Figure 4.10). Landscape ecology has been more intensively studied in the human-dominated environments of Europe than in North America, where research in the past has emphasized single habitat types.

Landscape ecology is important to the protection of biological diversity because many species are not confined to a single habitat, but move between habitats or live on borders where two habitats meet. For these species, the patterns of habitat types on a regional scale are of critical importance. The presence and density of many species may be affected by the size of habitat patches and their degree of linkage. For example, the population size of a rare animal species will be different in two 100 ha parks, one with an alternating checkerboard of 100 patches of field and forest, each 1 ha in

4.10 Four renditions of landscape types are shown in which interacting ecosystems or other land uses form repetitive patterns. The discipline of landscape ecology focuses on such interactions rather than on a single habitat type. (After Zonneveld and Forman 1990.)

(A) Scattered patch landscapes

Open clearings in a forest

Groves of trees in a field

(B) Network landscapes

Network of roads in a large plantation

Riparian network of rivers and tributaries in a forest

(C) Interdigitated landscapes

Tributary streams running into a lake

Shifting forest—grassland borders

(D) Checkerboard landscapes

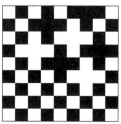

Farmland under cultivation for different crops

Lots in a residential development

area, the other with a checkerboard of 4 patches, each 25 ha in area. These alternative landscape patterns may have very different effects on microclimate (wind, temperature, humidity, and light), pest outbreaks, and animal movement patterns.

To increase the number and diversity of animals, wildlife managers often attempt to create the greatest amount of landscape variation possible within the confines of their management unit. Fields and meadows are created and maintained, small thickets are encouraged, fruit trees and crops are planted on a small scale, small patches of forests are periodically cut, small ponds and dams are developed, and numerous trails and dirt roads meander across and along all of the patches. The result is a park transformed into a mass of edges, where transition zones abound. In one textbook of wildlife management, managers are advised to "develop as much edge as possible" because "wildlife is a product of the places where two habitats meet" (Yoakum and Dasmann 1971).

The conservation biologist's goal, however, is not just to include as many species as possible within nature reserves, but also to protect those species most in danger of extinction as a result of human activity. Small reserves broken up into many small habitat units within a compressed landscape may have a large number of species, but these are likely to be principally "weedy" species—species that depend on human disturbance—and nonnative species. A park that contains the maximum amount of edge may lack many rare species that inhabit only large blocks of undisturbed habitat.

To remedy this localized approach, biological diversity needs to be managed on the regional landscape level, at which the size of the landscape units—such as entire watersheds or groups of hills—more closely approximates the natural units prior to human disturbance (Grumbine 1994b; Noss and Cooperrider 1994). An alternative to creating a miniature landscape of contrasting habitats on a small scale is to link all parks in an area in a regional plan so that larger habitat units can be created. Some of these larger habitat units would then be large enough to protect rare species that are not able to tolerate human disturbance.

Managing Protected Areas

Once a protected area has been legally established, it must be effectively managed if biological diversity is to be maintained. The conventional wisdom that "nature knows best" and that there is a "balance of nature" leads some people to the conclusion that biodiversity is best served when there is no human intervention. The reality is often very different: In many cases, humans have already modified the environment so much that the remaining species and communities need human intervention in order to survive (Speller-berg 1994; Sutherland and Hill 1995; Halverson and Davis 1996).

The world is littered with "paper parks" that have been created by government decree but not effectively managed on the ground. These parks have gradually—or sometimes rapidly—lost species, and their habitat quality has been degraded. In some countries, people do not hesitate to farm, log, and mine in protected areas because they believe that government land is owned by "every-one"; "anybody" can take whatever they want and "nobody" is willing to intervene. The crucial point is that parks must sometimes be actively managed to prevent deterioration. In addition, the most effective park management usually occurs when managers have the benefit of information from a research program and have funds available to implement management plans. In some countries, par-ticularly European countries such as the United Kingdom, the habi-tats of interest, such as forests, meadows, and hedges, have been formed from hundreds and even thousands of years of human activity. These habitats support high species diversity as a result of traditional land-management practices, which must be maintained if the species are to persist.

It is also true that sometimes the best management involves doing nothing: Management activities are sometimes ineffective or even detrimental. For example, active management to promote the abundance of game species, such as deer, has frequently involved eliminating top predators, such as wolves and cougars. Removal of top predators can result in an explosion of game populations (and, incidentally, rodents). The result is overgrazing, habitat degrada-tion, and a collapse of the animal and plant communities. Overen-thusiastic park managers who remove fallen trees and underbrush to "improve" a park's appearance may unwittingly remove a criti-cal resource needed by certain animal species for nesting and over-wintering. In many parks, fire is part of the ecology of the area. Attempts to suppress fire completely are expensive and unnatural, eventually leading to massive, uncontrolled fires such as those that occurred in Yellowstone National Park in 1988.

Many good examples of park management come from Britain, where there is a history of scientists and volunteers successfully mon-itoring and managing small reserves, such as the Monks Wood and Castle Hill Nature Reserves (Peterken 1996). At these sites, the effects of different grazing methods (sheep vs. cattle, light vs. heavy graz-ing) on populations of wildflowers, butterflies, and birds are closely followed. In a symposium called *The Scientific Management of Animal and Plant Communities for Conservation* (Duffey and Watt 1971), Michael Morris of Monks Wood concluded, "There is no inherently right or wrong way to manage a nature reserve . . . the aptness of any method of management must be related to the objects of management for any particular site. . . . Only when objects of management have been formulated can results of scientific management be applied."

Many government agencies and conservation organizations have clearly articulated the protection of rare and endangered species as one of their top priorities. These priorities are often outlined in mission statements, which allow managers to defend their actions. For example, at the Cape Cod National Seashore in Massachusetts, protecting tern nesting habitat has been given priority over the use of the area by off-road vehicles and the "right" of sport fishermen to fish (Figure 4.11). A "hands-off" policy by park managers, ostensibly allowing nature to take its course, would result in the rapid destruction of the tern colonies.

An important aspect of park management involves establishing a monitoring program of key components related to biological diversity, such as water levels in ponds, the number of individuals of rare and endangered species, the density of herbs, shrubs, and trees, and the dates migratory animals arrive at and leave the park. The exact types of information gathered depend on the goals of park management. Not only does monitoring allow managers to determine the health of the parks, it can suggest which management practices are working and which are not. With the right infor-

4.11 Tern nesting habitat in the Cape Cod National Seashore and at nearby beaches is extremely vulnerable to the "wear and tear" that is inevitable in heavily visited recreation areas. Management is needed to reduce the impact of such human activity on the birds. (Photograph by David C. Twichell, Manomet Bird Observatory.)

mation in hand, managers may be able to adjust management practices to increase the chances of success.

The World Conservation Monitoring Centre and UNESCO conducted a survey in 1990 of 89 World Heritage Sites to identify their management problems (WRI/IUCN/UNEP 1992). The most serious management problems in Oceania (Australia, New Zealand, and the Pacific Islands) were introduced plant species, while illegal wildlife harvesting, fire, grazing, and cultivation were the major threats in both South America and Africa. Inadequate park management was a particular problem in the developing countries of Africa, Asia, and South America. The greatest threats faced by parks in industrialized countries were internal and external threats associated with economic activities such as mining, logging, agriculture, and water projects. Although these general patterns give an overview, any single park has its own unique problems, such as the illegal logging and hunting that occur in many Central American parks and the vast numbers of tourists who crowd into U.S. and European parks in the summer.

Habitat management

A park may have to be aggressively managed to ensure that the full range of species and habitats are protected and maintained (Richards et al. 1999). Many species occupy only a specific habitat or a specific successional stage of a habitat. When land is set aside as a protected area, the pattern of disturbance and human use may change so markedly that many species previously found on the site fail to persist. Natural disturbances, including fires, grazing, and tree falls, are key elements in the presence of certain rare species. In small parks, the full range of successional stages may not be present, and many species may be missing for this reason; for example, in an isolated park dominated by old-growth trees, species characteristic of the early successional herb and shrub stage may be missing. Park managers sometimes must actively manage sites to ensure that all successional stages are present. One common way to do this is to periodically set localized, controlled fires in grasslands, shrublands, and forests to reinitiate the successional process. For example, many of the unique wildflowers of Nantucket Island, which is located off the coast of Massachusetts, are found in the scenic heathland areas. These heathlands were previously maintained by sheep grazing; they must now be burned every few years to prevent scrub oak forest from taking over the area and shading out the wildflowers (Figure 4.12A). In some wildlife sanctuaries, managers maintain open fields by mowing or grazing livestock. The type and timing of management can also help in removing exotic species that are invading at the expense of native species (Weiss 1999). In other sit-

(A)

4.12 Conservation management: intervention versus leave-it-alone. (A) Heathland in protected areas of Nantucket Island is burned on a regular basis in order to maintain the open vegetation habitat and to protect wildflowers and other rare species. (Photograph by Jackie Sones, Massachusetts Audubon Society.) (B) Sometimes management involves keeping human disturbance to an absolute minimum. This old-growth stand in the Olympic National Forest, Washington, is the result of many years of solitude. (Photograph by Thomas Kitchin/ Tom Stack and Associates.)

(B)

uations, parts of protected areas must be carefully managed to minimize human disturbance and to provide the conditions required by old-growth species (Figure 4.12B). For example, certain ground beetle species are found only in mature stands of boreal forest and disappear from lands managed under a system of clear-cut harvesting (Niemelå et al. 1993).

Wetlands. Wetlands management is a particularly crucial issue. The maintenance of wetlands is necessary to preserve populations of water birds, fish, amphibians, aquatic plants, and a host of other species (Moyle and Leidy 1992). Wetlands are often interconnected, so a decision affecting water levels and quality in one place has repercussions in other areas. Yet parks may become direct competitors for water resources with irrigation projects, flood control schemes, and hydroelectric dams in places such as the floodplains of India and the Florida Everglades (Holloway 1994). Park managers may have to become politically sophisticated and effective at public relations to ensure that the wetlands under their supervision continue to receive the clean water they need to survive.

Rare species. The need for habitat management to maintain populations of rare species is illustrated by the example of Crystal Fen in northern Maine, recognized for its numerous rare plant species (Jacobson et al. 1991). An apparent drying of the fen (a type of wet meadow) and an increase in woody vegetation, which were attributed by some biologists to the construction of a railroad in 1893 and a drainage ditch in 1937, led to concern that the fen community might be lost. Subsequent studies using aerial photography, vegetation history, and dated fossil remains from peat layers collectively showed that the construction of the railroad bed had allowed the wetland to *expand* in area by impeding drainage. The fen had also increased in area following fires that were started by cinder-producing locomotives. Today, the large area of fen in which rare plants occur is primarily a product of human activity. The construction of the drainage ditch and the decrease in fires following the change to diesel-powered locomotives are allowing the vegetation to return to its original state. If the goal is to maintain the current extent of the fen and its populations of rare species, management practices such as periodic burning, removing woody plants, and manipulating drainage patterns are necessary.

Keystone resources. In managing parks, attempts should be made to preserve and maintain keystone resources on which many species depend (see Chapter 1). If it is not possible to keep these keystone resources intact, attempts can be made to reconstruct them. For example, artificial pools could be built in streambeds to provide replacement water supplies. Keystone resources and keystone species could conceivably be enhanced in managed conservation areas to increase the populations of species whose numbers have declined. For example, by planting fruit trees and building an artificial pond, it might be possible to maintain vertebrate species in a smaller conservation area at higher densities than would be predicted based on studies of species distribution in undisturbed habitat. Another example is the provision of nesting boxes for birds as a substitute resource when few dead trees with nesting cavities are available. In this way a viable population of a rare species might be established, whereas without such interventions the population size might be too small for it to persist. In each case, a balance must be struck between establishing nature reserves free from human influence and creating seminatural gardens in which the plants and animals are dependent on people.

Park management and people

Human use of the landscape is a reality that must be addressed in park design. People have been a part of virtually all the world's ecosystems for thousands of years, and excluding humans from nature reserves can have unforeseen consequences (Kramer et al. 1997). For example, a savannah protected from fires set by people may change to forest, with a subsequent loss of the savannah species. However, excluding local people from protected areas may be the only option when resources are being overharvested to the point where the integrity of the biological communities is being threatened. Such a situation could result from overgrazing by cattle, excessive collection of fuelwood, or hunting with guns. It is better if compromises can be found before this point is reached.

The use of parks by local people and outside visitors must be a central part of any management plan, in both developed and developing countries (Wells and Brandon 1992; Western et al. 1994; Kothari et al. 1996). People who have traditionally used products from inside a nature reserve and are suddenly not allowed to enter the area will suffer from their loss of access to the basic resources that they need to stay alive. They will be understandably angry and frustrated, and people in such a position are unlikely to be strong supporters of conservation. Many parks flourish or are destroyed depending on the degree of support, neglect, hostility, or exploitation they receive from the people who use them. If the purpose of a protected area is explained to local residents, and if most residents agree with the objectives and respect the rules of the park, then the area may better maintain its natural communities. In the most positive scenario, local people become involved in park management and planning, are trained and employed by the park authority, and benefit from the protection of biodiversity and regulation of activity within the park. At the other extreme, if there is a history of bad relations and mistrust between local people and the government, or if the purpose of the park is not explained adequately, the local people may reject the park concept and ignore park regulations (Brandon et al. 1998). In this case, the local people will come into conflict with park personnel, to the detriment of the park.

There is now increasing recognition that the involvement of local people is the crucial missing element in many conservation management strategies. "Top-down" strategies, in which governments try to impose conservation plans, need to be integrated with "bottom-up" programs, in which villages and other local groups are able to formulate and reach their own development goals (Clay 1991). As explained by Lewis et al. (1990):

> If any lesson can be learned from past failures of conservation in Africa, it is that conservation implemented solely by government for the presumed benefit of its people will probably have limited success, especially in countries with weakened economies. Instead, conservation for the people and by the people with a largely service and supervisory role delegated to the government could foster a more cooperative relationship between government and the residents living with the resource. This might reduce the costs of law enforcement and increase revenues available to other aspects of wildlife management, which could help support the needs of conservation as well as those of the immediate community. Such an approach would have the added advantage of restoring to local residents a greater sense of traditional ownership and responsibility for this resource. Convincing proof that such a partnership is possible has yet to be demonstrated and has therefore been more theoretical than practical.

The United Nations Educational, Scientific, and Cultural Organization (UNESCO) has pioneered one such approach with its Man and the Biosphere (MAB) Program. This program has designated hundreds of Biosphere Reserves worldwide in an attempt to integrate human activities, research, protection of the natural environment, and tourism at a single location (Figure 4.13A; Batisse 1997). The Biosphere Reserve concept involves a core area in which biological communities and ecosystems are strictly protected, surrounded by a buffer zone in which traditional human activities—such as collection of thatch, medicinal plants, and small fuelwood—are monitored, and nondestructive research is conducted. Surrounding the buffer zone is a transitional zone in which some forms of sustainable development (such as small-scale farming) are allowed, along with some extraction of natural resources (such as selective logging) and experimental research (Figure 4.13B). This general strategy of sur-

(A)

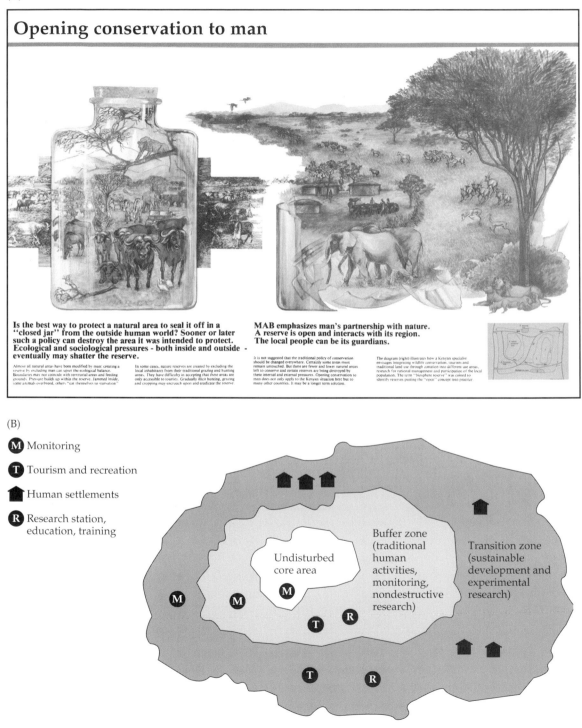

Opening conservation to man

Is the best way to protect a natural area to seal it off in a "closed jar" from the outside human world? Sooner or later such a policy can destroy the area it was intended to protect. Ecological and sociological pressures - both inside and outside - eventually may shatter the reserve.

Almost all natural areas have been modified by man: creating a reserve by excluding man can upset the ecological balance. Boundaries may not coincide with territorial areas and feeding grounds. Pressure builds up within the reserve. Jammed inside, some animals overbreed, others "eat themselves to starvation."

In some cases, nature reserves are created by excluding the local inhabitants from their traditional grazing and hunting areas. They have difficulty in accepting that those areas are only accessible to tourists. Gradually illicit hunting, grazing and cropping may encroach upon and eradicate the reserve.

MAB emphasizes man's partnership with nature. A reserve is open and interacts with its region. The local people can be its guardians.

It is not suggested that the traditional policy of conservation should be changed everywhere. Certainly some areas must remain untouched. But there are fewer and fewer natural areas left to conserve and certain reserves are being destroyed by these internal and external pressures. Opening conservation to man does not only apply to the Kenyan situation here but to many other countries. It may be a longer term solution.

The diagram (right) illustrates how a Kenyan specialist envisages integrating wildlife conservation, tourism and traditional land use through zonation into different use areas, research for rational management and participation of the local population. The term "biosphere reserve" was coined to identify reserves putting the "open" concept into practice.

(B)

M Monitoring

T Tourism and recreation

🏠 Human settlements

R Research station, education, training

Undisturbed core area **M**

Buffer zone (traditional human activities, monitoring, nondestructive research)

Transition zone (sustainable development and experimental research)

M **M**

T **R**

T **R**

◀ **4.13** Past management policies have often attempted to protect natural areas by sealing them off from outside influences. (A) As this UNESCO poster illustrates, the MAB program is an attempt to integrate the needs and cultures of local people in park planning and protection. (Poster from "Ecology in Action: An Exhibit," UNESCO, Paris, 1981.) (B) The general pattern of an MAB reserve includes a core protected area, surrounded by a buffer zone where traditional human activities are monitored and managed, and research is carried out; this, in turn, is surrounded by a transition zone where sustainable development and experimental research take place.

rounding core conservation areas with buffer and transitional zones can have several desirable effects. First, the arrangement may encourage local people to support the goals of the protected area. Second, certain desirable features of the landscape created by human use may be maintained. And third, the buffer zones may facilitate animal dispersal and gene flow between highly protected core conservation areas and human-dominated transitional and unprotected areas.

Managers of marine parks are also attempting to find the balance between the need for conservation and development (Scott 1999). In many marine parks, certain areas are closed to fishing, in order to allow populations of fish and other organisms to rebuild (Figure 4.14). The hope is that individuals in these protected areas will eventually disperse to managed areas and rebuild damaged and destroyed populations. In other parks, competing uses can be balanced by zoning, in which certain activities have priority. For example, at the Ningaloo Marine Park, off the western coast of Australia, areas are assigned to one of the three management categories: **sanctuary zones,** which minimize human impact and permit only view-

4.14 Large reef fish had been overharvested at Apo Island in the Philippines and were rarely seen. (A) In response to the overharvesting, a reserve was set up (shaded area) on the eastern side of the island. Fishing continued at a nonreserve area on the western side of the island. A censusing study measured the number of large reef fish in each site (six underwater census areas are shown for each). (B) Resulting data show that after the eastern part of the island was protected as a marine reserve, the number of fish observed in the unfished reserve increased substantially; the number of fish in the nonreserve area did not increase because the fish were still being intensively harvested. (After Russ and Alcala 1996.)

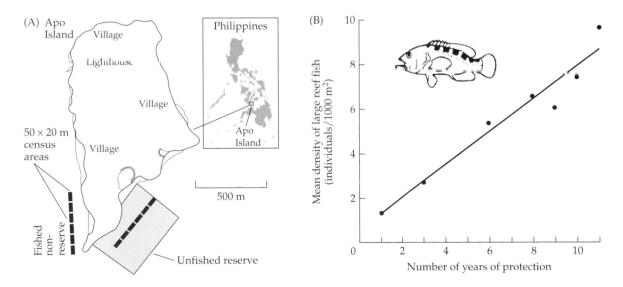

ing of marine life; **recreation zones,** which may include activities such as swimming, boating, and recreational fishing; and **general use zones,** which allow commercial and recreational fishing.

For park management to be effective there must be sufficient numbers of well-equipped, properly trained and motivated park personnel who are willing to carry out park policy. In many areas of the world, particularly in developing countries, protected areas are understaffed, and they lack the vehicles and equipment to patrol remote areas of the park (Peres and Terborgh 1995). The importance of sufficient personnel and equipment should not be underestimated: In areas of Panama, for instance, the abundance of large mammals and the seed dispersal services they provide are directly related to the frequency of anti-poaching patrols by park guards (Wright et al. 2000). It is an irony of our world that vast sums are spent on captive breeding and conservation programs by zoos and conservation organizations in the developed world, while the biologically rich parks of so many developing countries languish for lack of resources. In many cases, the annual management costs for endangered species and habitats are a bargain compared with the large costs of conservation efforts to save species on the verge of extinction or ecosystems on the verge of collapse (Wilcove and Chen 1998).

Outside Protected Areas

A crucial element in conservation strategies must be the protection of biological diversity outside as well as inside protected areas. As stated by David Western, former Director of the Kenya Wildlife Service (1989), "If we can't save nature outside protected areas, not much will survive inside." A danger of relying on parks and reserves alone is that this strategy can create a "siege mentality," in which species and communities inside the parks are rigorously protected while those outside are left to be freely exploited. If the areas surrounding parks are degraded, however, biological diversity within the parks will decline as well, with the loss of species being most severe in small parks (Table 4.5). This decline will occur because many species must migrate across park boundaries to gain access to resources beyond those that the park itself can provide. Also, the number of individuals of any one species contained within park boundaries may be lower than the minimum viable population size.

African wildlife outside parks

The abundance of African wildlife outside its parks helps illustrate the importance of maintaining biological diversity outside pro-

TABLE 4.5 Predicted decline of large herbivore species in African parks if areas outside parks exclude wildlife

National park	Area (× 1000 ha)	No. of species currently in park	If outside areas exclude wildlife[a]	
			No. of species that will persist	% of species that will persist
Serengeti, Tanzania	1450	31	30	97
Mara, Kenya	181	29	22	76
Meru, Kenya	102	26	20	77
Amboseli, Kenya	39	24	18	75
Samburu, Kenya	30	25	17	68
Nairobi, Kenya	11	21	11	52

Source: Data from Western and Ssemakula 1981.
[a]Due to agriculture, hunting, herding, or other human activities.

tected areas. East African countries such as Kenya are famous for the large wildlife species found in their national parks, which are the basis of a valuable ecotourism industry. Despite the fame of these parks, however, about three-fourths of Kenya's 2 million large animals live in areas outside them, often sharing rangeland with domestic cattle (Western 1989). The rangelands of Kenya occupy 700,000 km², or about 40% of the country. Among the well-known species found predominantly outside the parks are giraffes (89%), impalas (72%), Grevy's zebras (99%), oryx (73%), and ostriches (92%). Only rhinoceroses, elephants, and wildebeest are found predominantly inside the parks; rhinos and elephants are concentrated there because poachers seeking ivory, horns, and hides have virtually eliminated external populations of these animals. The large herbivores found in the parks often graze seasonally outside of them; many of these species would be unable to persist if fencing restricted them to the limits of the parks, or if external poaching and agricultural development necessitated their restriction to the parks' interiors.

Within Kenya, Zambia, Zimbabwe, and other neighboring countries, a new government policy is being created to allow rural communities and landowners to profit directly from the presence of large game animals on unprotected land (Lewis 1995; Western 1997). With assistance from international donor agencies, local ecotourist businesses—involving hiking, photography, canoeing, and horseback safaris—are being established. Where the land is adequately stocked with animals, trophy hunting is sometimes allowed for high fees. This revenue is shared between the local communities and the national government; in some cases, meat and hides from hunting expeditions are being sold for additional shared revenue. While the programs are in operation, communities receiving such revenue have a strong incentive to protect wildlife and prevent poaching. The most well-known programs are CAMPFIRE in Zimbabwe and ADMADE in Zambia (Getz et al. 1999). The programs are apparently successful at combining conservation and community development, but they have also been criticized by animal rights groups because they allow trophy hunting. Further, the programs depend on continuing subsidies from outside donor agencies, with only a small percentage of funding actually reaching the village level. When outside subsidies cease, the programs often end, indicating the weakness of the local economy, the instability of the ecotourism industry, and ineffective government policies.

Strategies for success

More than 90% of the world's land will remain outside of protected areas, according to even the most optimistic predictions. As the African wildlife example shows, strategies for reconciling human needs and conservation interests in these unprotected areas are critical to the success of conservation plans (Redford and Richter 1999). Because the majority of the land area in most countries will never be in protected areas, numerous rare species will inevitably occur outside protected areas. In the United States, more than one-third of the species listed under the U.S. Endangered Species Act are found exclusively on private land, and more than half of the listed species are found predominantly on private land. Conservation strategies in which private landowners are educated and encouraged to protect rare species, and even compensated in some way, are obviously the key to the long-term survival of many species. Government endangered species programs in many countries inform road builders and developers of the locations of rare species and assist them in modifying their plans to avoid damage to the sites. Forests that are selectively logged on a long cutting cycle or are used for traditional shifting cultivation by a low density of farmers can maintain a considerable percentage of their original biota (Thiollay 1992; Chapman et al. 2000). In Malaysia, the majority of bird species are still found in rain forests 25 years after selective logging began because undisturbed forest is available nearby to act as a source of colonists. Considerable biological diversity can also be maintained in traditional agricultural systems and forest plantations. Many bird species of wetlands can readily adapt to feeding in flooded rice fields. One well-studied example is the shade coffee plantation, in which coffee is grown under many species of shade trees (Roberts et al. 2000). These coffee plantations have a diversity of birds and insects that is sometimes comparable to adjacent natural forest.

Native species can often continue to live in unprotected areas when those areas are set aside or managed for some other purpose that is not harmful to the ecosystem. Security zones surrounding government installations are some of the most outstanding natural areas in the world. In the United States, excellent examples of natural habitat occur on the sites of military reservations such as Fort Bragg in North Carolina; nuclear processing facilities such as the Savannah River site in South Carolina; and watersheds adjacent to metropolitan water supplies such as the Quabbin Reservoir in Massachusetts. Although dams, reservoirs, canals, dredging operations, port facilities, and coastal development destroy and damage aquatic communities, some species are capable of adapting to the altered conditions, particularly when the water itself is not polluted. In estuaries and seas managed for commercial fisheries, many of the native species remain, because commercial and noncommercial species alike require that the chemical and physical environments are undamaged.

Other areas that are not protected by law may retain biological diversity because their human population density and degree of utilization are typically very low. Border areas such as the demilitarized zone between North and South Korea often have an abundance of wildlife because they remain undeveloped and unpopulated. Mountain areas are often too steep and inaccessible for development. These areas are frequently managed by governments as watersheds for their value in producing a steady supply of water and preventing flooding, yet they also harbor natural communities. Likewise, desert communities may be at less risk than other unprotected communities because desert regions are considered marginal for human habitation and use. In many parts of the world, wealthy individuals have acquired large tracts of land for their personal estates and for private hunting. These private estates frequently are used at a very low intensity, often in a deliberate attempt by the landowner to maintain large wildlife populations. Some estates in Europe that have been owned and protected for hundreds of years by royal families have preserved unique old-growth forests.

Large parcels of government-owned land in many countries are designated for multiple uses. In the past, such uses have included logging, mining, grazing, wildlife management, and recreation. Increasingly, multiple-use lands are also being valued and managed for their ability to protect species, biological communities, and ecosystems (Figure 4.15; Noss and Cooperrider 1994; Szaro and Johnston 1996; Hunter 1999; Donahue 1999). For instance, the protection of biological diversity is being incorporated into "ecological forestry," which has been developed for the Pacific Northwest (Kohm and Franklin 1997; Carey 2000). This method involves leav-

4.15 Longleaf pine forest in the southeastern United States, including areas of South Carolina, North Carolina, and Georgia, has traditionally been managed by the U.S. Forest Service for timber production, but is now also being managed to protect the endangered red-cockaded woodpecker (*Picoides borealis*) and the groves of old trees that the birds need for breeding. (A) Heavily logged areas lack older trees with the nesting holes that the woodpecker requires; in managed forests, artificial nesting holes are drilled in the trees. (B) Here a young red-cockaded woodpecker leaves the nest for its first flight. (Photographs © Derrick Hamrick.)

ing a few large live trees, standing dead trees, and some fallen trees during clear-cutting operations to provide some structural complexity and to serve as habitat for animal species during forest regrowth. Also, when forests are logged in larger blocks and large residual patches are untouched, additional rare species of wildlife can be maintained (Chapin et al. 1998).

Expanding the objectives of multiple-use lands to include protection of biological diversity extends even to the military: Military commanders are including wildlife protection as a management goal for the lands under their control. This is mandated by law, and is often valuable in maintaining good public relations. In addition, conservation biologists are using laws and court systems to halt government-approved activities on public lands that threaten the survival of endangered species. These and other examples show that conservation can be an important management consideration even on lands outside of protected areas.

Ecosystem management

Many land managers around the world are expanding their goals to include the health of ecosystems. The developing concept of **ecosystem management** is described by Grumbine (1994a) as follows: "Ecosystem management integrates scientific knowledge of ecological relationships within a complex sociopolitical and values framework toward the general goal of protecting native ecosystem integrity over the long term." Resource managers increasingly are being urged to expand their traditional emphasis on the maximum production of goods (such as volume of timber harvested) and services (such as the number of visitors to parks) and instead take a broader perspective that includes the conservation of biological diversity and the protection of ecosystem processes (Christenson et al. 1996; Yaffee et al. 1996; Yaffee 1999; Sexton et al. 1999; Poiani et al. 2000). Rather than each government agency, private conservation organization, business, or landowner acting in isolation, ecosystem management envisions them cooperating to achieve common objectives. For example, in a large forested watershed along a coast, ecosystem management would link all owners and users from the tops of the hills to the seashore, including foresters, farmers, conservation biologists, business groups, townspeople, and the fishing industry (Costanza et al. 1998; Figure 4.16). Not all ecologists have accepted the paradigm of ecosystem management, however. Some consider ecosystem management unlikely to change the human-oriented management practices that lead to the overexploitation of natural resources (Stanley 1995). Despite the lack of universal agreement among ecologists, it is clear that the concept of ecosystem management linked to the practice of conservation biology is being strongly embraced by government, business, and conservation groups.

Important themes in ecosystem management include:

1. Seeking connections among all levels and scales in the ecosystem hierarchy; for example, from the individual organism to the species to the community to the ecosystem.
2. Ensuring viable populations of all species, representative examples of all biological communities and successional stages, and healthy ecosystem functions.
3. Monitoring significant components of the ecosystem (numbers of individuals of significant species, vegetation cover, water quality, etc.), gathering the needed data, and then using the results to adjust management practices in an adaptive manner (sometimes called **adaptive management**).
4. Changing the rigid policies and practices of land management agencies, which often result in a piecemeal approach. Instead, interagency cooperation and integration at the local, regional, national, and international levels, and cooperation and commu-

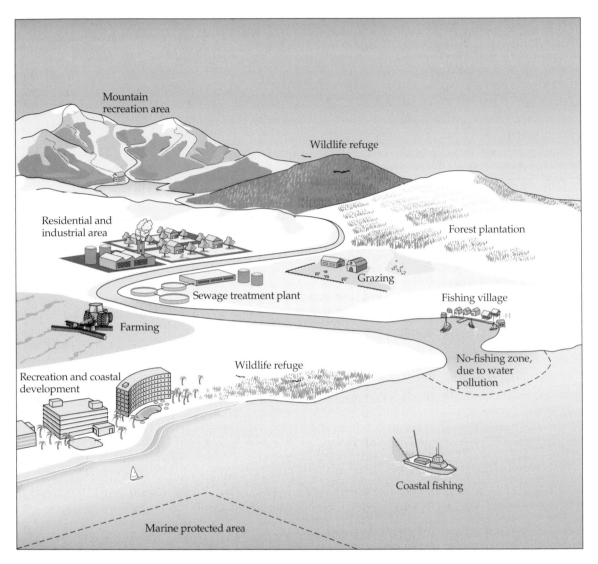

Mountain recreation area

Wildlife refuge

Residential and industrial area

Forest plantation

Sewage treatment plant

Grazing

Fishing village

Farming

No-fishing zone, due to water pollution

Recreation and coastal development

Wildlife refuge

Coastal fishing

Marine protected area

4.16 Ecosystem management involves joining together all of the stakeholders who affect a large ecosystem and receive benefits from it. In this case, a watershed needs to be managed for a wide variety of purposes, many of which influence each other. (After Miller 1996.)

nication encouraged between public agencies and private organizations.

5. Minimizing external threats to the ecosystem and maximizing sustainable benefits derived from it.

6. Recognizing that humans are a part of ecosystems, and that human needs and values influence management goals.

One example of ecosystem management is the Malpai Borderlands Group, a cooperative enterprise of ranchers and local landowners who promote collaboration with government agencies and conservation organizations such as The Nature Conservancy. The group is working to develop a network of cooperation across nearly 400,000 ha of unique desert habitat along the Arizona and New Mexico border. This rugged region of isolated mountains, or "sky islands," is one of the richest biological areas in the United States, supporting Mexican jaguars and numerous rare and endangered species. Priorities include controlling exotic grasses through fire management, controlling residential development through the use of conservation easements, and incorporating research into management plans.

A logical extension of ecosystem management is **bioregional management**, which often focuses on a single large ecosystem, such as the Caribbean Sea, the Great Barrier Reef of Australia, or the protected areas of Central America. A bioregional approach is particularly appropriate where there is a single, continuous, large ecosystem that crosses international boundaries. For example, 15 countries participate in the Mediterranean Action Plan to maintain the health of this enclosed sea (Figure 4.17).

Restoration Ecology

An important opportunity for conservation biologists is the chance to participate in the restoration of damaged or degraded ecosystems

4.17 The countries participating in the Mediterranean Action Plan cooperate in monitoring and controlling pollution and coordinating their protected areas. Major protected areas along the coast are shown as dots. (After Miller 1996.)

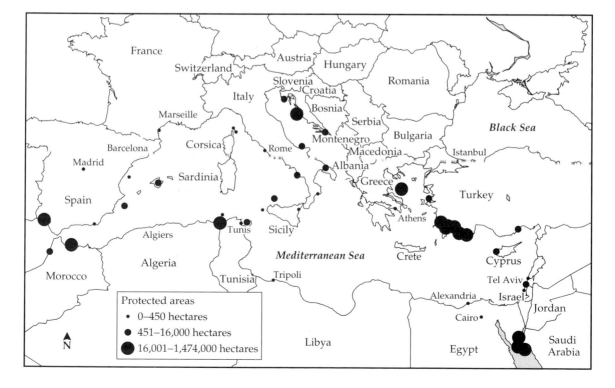

(Daily 1995). Rebuilding damaged ecosystems has great potential for enlarging and enhancing the current system of protected areas. **Ecological restoration** is defined as "the process of intentionally altering a site to establish a defined, indigenous, historic ecosystem. The goal of this process is to emulate the structure, function, diversity and dynamics of the specified ecosystem" (Society for Ecological Restoration 1991). **Restoration ecology** refers to the research and scientific study that investigates methods of carrying out these restorations. Restoration ecology has its origins in older applied technologies that restore ecosystem functions of known economic value: wetland replication to prevent flooding, mine site reclamation to prevent soil erosion, range management to ensure the production of grasses, and forest management for timber and amenity value (Gilbert and Anderson 1998). However, these technologies sometimes produce only simplified communities, or communities that cannot maintain themselves. With the emergence of biological diversity as an important societal concern, the reestablishment of species assemblages and entire communities has been included as a major goal in restoration plans. In many cases, businesses are required by law to restore habitats that they have damaged.

Ecosystems can be damaged by natural phenomena such as fires caused by lightning, volcanoes, and storms, but they typically recover their original biomass, community structure, and even a similar species composition through the process of succession. Some ecosystems damaged by human activity, however, are so degraded that their ability to recover is severely limited. Recovery is unlikely when the damaging agent is still present in the ecosystem. For example, restoration of degraded savannah woodlands in western Costa Rica and the western United States is not possible as long as the land continues to be overgrazed by introduced cattle; reduction of the grazing pressure is obviously the key starting point in restoration efforts (Fleischner 1994). Recovery is also unlikely when many of the original species have been eliminated over a large area, so that there is no source of colonists. For example, prairie species were eliminated from huge areas of the midwestern United States when the land was converted to agriculture. Even when an isolated patch of land is no longer cultivated, the original community does not become reestablished because there is no source of seeds or colonizing animals of the original species. In addition, recovery is unlikely when the physical environment has been so altered that the original species can no longer survive at the site; examples of this situation include mine sites, where the restoration of natural communities may be delayed by decades or even centuries due to the poor structure, heavy metal toxicity, and low nutrient status of the soil (Figure 4.18).

4.18 To speed the recovery of this devastated coal mine site in Wyoming, crews planted 120,000 shrubs. Mining sites often need a great deal of human help in order to recover even a semblance of biodiversity. (From Jordan et al. 1990; photograph courtesy of William R. Jordan III.)

In certain cases entirely new environments are created by human activity; examples include reservoirs, canals, landfills, and industrial sites. If these sites are neglected, they often become dominated by exotic and weedy species, resulting in biological communities that are unproductive, not typical of the surrounding areas, valueless from a conservation perspective, and aesthetically unappealing. If these sites are properly prepared and native species are reintroduced, aspects of the original communities can be successfully restored. New habitats are often deliberately created as part of a **mitigation** process to compensate for habitats damaged or destroyed elsewhere. The goal of these and other restorations efforts is to create habitats that are comparable to existing **reference sites**, in terms of ecosystem functions or species composition. Reference sites provide explicit goals for restoration, providing quantitative measures of a project's success (White and Walker 1997; Stephenson 1999; Kloor 2000).

Restoration ecology provides the theory and techniques to address various types of degraded ecosystems. Four main approaches are available in restoring biological communities and ecosystems (Figure 4.19; Bradshaw 1990; Cairns and Heckman 1996).

1. **No action** because restoration is too expensive, because previous attempts at restoration have failed, or because experience has shown that the ecosystem will recover on its own. The last

4.19 Degraded ecosystems have lost their structure (in terms of species and their interactions with the physical and biological environments) and their function (the accumulation of biomass and soil, water, and nutrient processes). Decisions must be made as to whether to restore, rehabilitate, or replace the degraded site, or whether the best course of action is no action. (After Bradshaw 1990.)

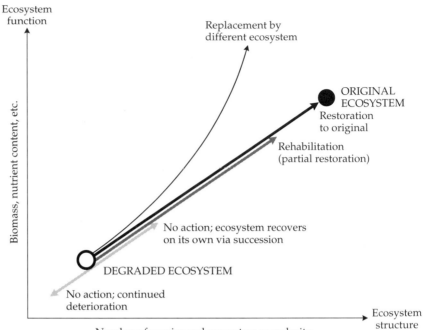

situation is typical of old fields in eastern North America, which return to forest within a few decades after being abandoned for agriculture.

2. **Replacement** of a degraded ecosystem with another productive ecosystem type; for example, replacing a highly degraded forest area with a productive pasture. Replacement at least establishes a biological community on a site, and it restores some ecosystem functions such as soil retention and flood control.

3. **Rehabilitation** to restore at least some of the ecosystem functions and some of the original, dominant species; for example, replacing a degraded forest with a tree plantation.

4. **Restoration** of the area to its original species composition and structure by an active program of reintroduction; in particular, by planting and seeding the original plant species. Factors degrading the habitat must be identified and reduced. Natural ecological processes must be reestablished to heal the system.

Civil engineers and others involved in major development projects deal with the restoration of degraded habitats in a practical, technical manner. Their goals are to find economical ways to permanently stabilize land surfaces, prevent soil erosion, make the site look better to neighbors and the general public, and if possible, restore the productive value of the land. Ecologists contribute to these restoration efforts by developing ways to restore the original communities in terms of species diversity, species composition, vegetation structure, and ecosystem function. Practitioners of restoration ecology must have a clear grasp of how natural systems work and what methods of restoration are feasible. To be practical, restoration ecology must also consider the speed of restoration, the cost, the reliability of the results, and the ability of the final community to persist with little or no further maintenance. Considerations such as the cost and availability of seeds, when to water plants, how much fertilizer to add, and how to prepare the surface soil may become paramount in determining the success of a project. Dealing with such practical details has not generally been attractive to academic biologists in the past, but they must be dealt with in restoration ecology.

Restoration ecology is valuable to the science of ecology because it provides a test of how well we understand biological communities by challenging us to reassemble them from their component parts. As Bradshaw (1990) has said, "Ecologists working in the field of ecosystem restoration are in the construction business, and like their engineering colleagues, can soon discover if their theory is correct by whether the airplane falls out of the sky, the bridge collapses, or the ecosystem fails to flourish." In this sense, restoration ecology can be viewed as an experimental methodology that complements existing programs of basic research on intact systems. Restoration ecology provides an opportunity to completely reassemble communities in different ways, to see how well they function, and to test ideas on a larger scale than would otherwise be possible (Dobson 1997a).

Restoration ecology in practice

Efforts to restore degraded terrestrial communities have emphasized the reestablishment of the original plant community. This emphasis is appropriate because the plant community typically contains the majority of the biomass and provides a structure for the rest of the community. However, more attention needs to be devoted to the other major components of the community. Mycorrhizal fungi and bacteria play a vital role in decomposition of organic matter and nutrient cycling (Miller 1990); soil invertebrates are important in creating soil structure; herbivorous animals are important in reducing plant competition and maintaining species diversity; and many vertebrates have vital functions as seed dispersers, insect predators, and soil diggers. Many of these nonplant species can be transferred to a restored site in sod samples; large animals and aboveground invertebrates may have to be deliberately caught in sufficient numbers and then released onto restored sites. If an area is going to be destroyed and then restored later, as might occur during strip mining, the top layer of soil, which contains the majority of buried seeds, soil invertebrates, and other soil organisms, could be carefully removed and stored for later use in restoration efforts.

Main Candidates for Ecological Restoration

Efforts to restore ecological communities have focused extensively on wetlands, lakes, urban areas, prairies, and forests. These environments have suffered severe alteration from human activities and are good candidates for restoration work.

Wetlands

Some of the most extensive work on restoration has been done on wetlands, including swamps, marshes, streams, small ponds, and rivers (Galatowitsch and Van der Valk 1996; Zedler 1996; Karr and Chu 1998). Wetlands are often filled in or damaged, because their importance in flood control, maintenance of water quality, and preservation of biological communities is either not known or not appreciated. More than half of the original wetlands in the United States have already been filled, and in heavily populated states such as California, over 90% have been lost (Cairns and Heckman 1996). Because of the U.S. government policy of "no net loss of wetlands," projects that fill in and damage wetlands must repair them or create new wetlands to compensate for those damaged beyond repair. The focus of these efforts has been on recreating the original hydrology of the area, followed by planting native species. In practice, such efforts to restore wetlands often do not closely match the biological communities or hydrological characteristics of reference sites. The subtleties of species composition, water movement, and the structure of soils, as well as the site history, are too difficult to match. However, the restored wetlands often do have some of the wetland plant species, they usually have some of the beneficial ecosystem characteristics such as flood control and pollution reduction, and they are frequently valuable for wildlife habitat. Further study and research of restoration methods will hopefully result in improvement in wetlands restoration.

An informative example of wetlands restoration comes from Japan, where parents, teachers, and children have built over 500 small dragonfly ponds next to schools and in public parks to provide habitat for dragonflies and other aquatic species (Primack et al. In press). These ponds focus an entire science and math curriculum: Dragonflies are an important symbol in Japanese culture, and are

(A)

(B)

(C)

4.20 (A) Children in Yokohama, Japan, are building a dragonfly pond next to their school. Activities involved in its construction include excavating the site, packing the bottom with clay, and reinforcing the banks with wooden posts; later, when the pond is completed, the children will fill it with aquatic plants and release dragonfly larvae. (B) A group of children and adults uses butterfly nets to check for the diversity and abundance of dragonflies at a Yokohama city pond during a city-sponsored maintenance day. They will also remove excess aquatic plants and exotic fish species from the pond. (C) This publicity poster, which exclaims "Let's build a dragonfly pond!" along with an extensive offering of brochures and practical manuals, is part of government efforts to interest schoolchildren and the general public to participate in programs to restore and enhance the environment. (Photographs courtesy of Seiwa Mori and Yokohama City Environmental Protection Bureau.)

useful as a starting point for teaching zoology, ecology, and principles of conservation. The schoolchildren are responsible for the regular weeding and maintenance of these "living laboratories," which helps them to feel an ownership of the project and to develop environmental awareness (Figure 4.20).

Lakes

Attempts to restore eutrophic lakes have not only provided practical management information but have also provided insight into the basic science of limnology (the study of the chemistry, biology, and physics of fresh water) that otherwise would not have been possible (MacKenzie 1996). One of the most dramatic and expensive examples of lake restoration is that of Lake Erie (Makarewicz and Bertram 1991). Lake Erie was the most polluted of the Great Lakes in the 1950s and 1960s, and was characterized by deteriorating water quality, extensive algal blooms, declining indigenous fish populations, the collapse of commercial fisheries, and oxygen depletion in deeper waters. To address this problem, the United States and Canadian governments have invested more than $7.5 billion since 1972 in wastewater treatment facilities, reducing the annual discharge of phosphorus into the lake from 15,260 tons in 1972 to 2449 tons in 1985. Once water quality began to improve in the mid-1970s and 1980s, stocks of the predatory fish, such as the walleye pike (*Stizostedion vitreum*), began to increase on their own and were also added to the lake by state agencies. These predatory fish were added because they eat small zooplankton-eating fish; with fewer small fish, the zooplankton increased, more algae were eaten, and the water quality further improved. Over the last few years, water clarity has increased substantially in the western part of the lake, probably due to invasion by zebra mussels, which filter algae out of the water.

There is even evidence of improvement in oxygen levels at the lower depths of the lake. Even though the lake will never return to its original condition because of the large number of exotic species present and the altered water chemistry, the combination of controls on water quality and the investment of billions of dollars has resulted in a significant degree of restoration in this large, highly managed ecosystem.

Urban areas

Highly visible restoration efforts are also taking place in many urban areas, reducing the intense human impact on ecosystems as well as enhancing the quality of life for city dwellers. Local citizen groups often welcome the opportunity to work with government agencies and conservation groups to restore degraded urban areas. Unattractive drainage canals in concrete culverts can be replaced with winding streams bordered with large rocks and planted with native wetland species. Vacant lots and neglected lands can be replanted with native shrubs, trees, and wildflowers. Gravel pits can be packed with soil and restored as ponds. These efforts have the additional benefits of fostering neighborhood pride, creating a sense of community, and enhancing local property values.

Restoring native communities to huge urban landfills presents one of the most unusual opportunities. In the United States, 150 million tons of trash are being buried in over 5000 active landfills each year. Some of these eyesores are now the focus of conservation efforts. When they have reached their maximum capacity, these landfills are usually capped by sheets of plastic and layers of clay to prevent toxic chemicals and pollutants from seeping out. If these sites are left alone, they are often colonized by weedy, exotic species. However, planting native shrubs and trees on the capped site attracts birds and mammals to bring in and disperse the seeds of a wide range of native species.

Prairies

Many small parcels of former agricultural land in North America have been restored as prairies (Kline and Howell 1990). Prairies represent ideal subjects for restoration work because they are species-rich, are home to many beautiful wildflowers, and can be established within a few years. Also, the technology used for prairie restoration is similar to that of gardening and agriculture and is well suited to incorporating volunteer labor.

Some of the most extensive research on the restoration of prairies has been carried out in Wisconsin, starting in the 1930s. A wide variety of techniques has been used in these prairie restoration attempts. The basic method involves a light site preparation by shallow plowing, burning, and raking if prairie species are present, or elimination of all vegetation by deeper plowing or herbicides if only exotics are present. Native plant species are then established by transplanting prairie sods that were obtained elsewhere, by planting individuals grown from seed, or by scattering prairie seed collected from the wild or from cultivated plants (Figure 4.21). The simplest method is gathering hay from a native prairie and sowing it on the prepared site.

One of the most ambitious and controversial proposed restoration schemes involves re-creating a prairie ecosystem, or "buffalo commons," on about 380,000 km^2 of the American Plains states, from the Dakotas to Texas and from Wyoming to Nebraska (Popper and Popper 1991; Mathews 1992). This land is currently used for environmentally damaging and often unprofitable agriculture and grazing, which are supported by government subsidies. The human population of this region is declining as farmers and townspeople go out of business and young people move away. From an ecological, sociological, and even an economic perspective, the best long-term use of much of the region might be a restored prairie ecosystem. The human population of the region could potentially stabilize around nondamaging core industries such as tourism, wildlife management, and low-level grazing, leaving only the best lands in agriculture. On a somewhat smaller scale, a project in Siberia is being planned to restore the original steppe grasslands ecosystem, including bison, wild horses, and other large grazers, on 160 km^2 of land.

4.21 (A) In the late 1930s, members of the Civilian Conservation Corps (one of the organizations created by President Franklin D. Roosevelt in order to boost employment during the Great Depression) participated in a University of Wisconsin project to restore the wild species of a Midwestern prairie. (B) The prairie as it looked 50 years later. (Photographs from the University of Wisconsin Arboretum and Archives.)

(A)

(B)

Tropical dry forests

Throughout the world, tropical dry forests have long been degraded by logging, grazing, fire, cultivation, and the collection of fuelwood (see Chapter 2). These lands often become degraded to the point at which they have few remaining trees and little value to local people. To reverse these trends, governments and local people are now protecting remaining forests and beginning to manage for restoration.

The tropical dry forests of Central America have long suffered from large-scale conversion to cattle ranches, logging, and farms. Only a few forest fragments remain. Even in these fragments, logging and hunting threaten the remaining species. This destruction has gone on largely unnoticed, as international scientific and public attention has focused on the more glamorous rain forests elsewhere. The American ecologist Daniel Janzen has been working with the government and a resident staff to restore 110,000 ha of tropical dry forest and adjacent rain forest in the Guanacaste Conservation Area (GCA) in northwestern Costa Rica (Janzen 1988, 1999, 2000). Restoration includes planting native trees, planting exotic trees to shade out introduced grass, eliminating human-caused fires, and banning logging and hunting. Livestock grazing is initially used to lower the abundance of grasses; grazing is then phased out as the forest invades through natural animal- and wind-borne seed dispersal. In just 15 years, this process has converted 60,000 ha of pastures to a species-rich, dense young forest (Figure 4.22). While this process reestablishes the dry forest ecosystem and benefits the adjacent rain forest to which animals of the dry forest seasonally migrate, it will require 200–500 years to regain the original forest structure.

An innovative aspect of this restoration is that all 130 members of the staff and administration of the GCA are Costa Ricans and reside in the area. The GCA offers training and advancement for its staff, educational opportunities for their children, and the best economic use of these marginal lands, which were formerly ranch and farm lands. GCA employees are selected from the local community, rather than spending scarce resources on imported consultants. A key element in the restoration plan is what has been termed *biocultural restoration*, meaning that the GCA teaches basic biology in the field to all students in grades four through six in the neighboring

4.22 (A) The Guanacaste National Park is an experiment in restoration ecology—an attempt to restore the native forest on these artificial grasslands, which were created by cattle grazing and frequent fires. (B) Eight years of fire suppression and controlled grazing have allowed a young forest to become established. (Photographs by C. R. Carroll.)

(A)

(B)

schools. This has created a community literate in conservation issues as well as a local viewpoint that the GCA offers something of value to all. Residents have come to see the GCA as if it were a large ranch producing "wildland resources" for the community, rather than an exclusionary "national park." The GCA is welcomed and viewed as being as important as the traditional agricultural countryside it replaced (Janzen 1999, 2000). Both the staff and neighboring residents have become strong supporters of the GCA.

Restoration Ecology and the Future of Conservation

Restoration ecology will play an increasingly valuable role in the conservation of biological communities if degraded lands and aquatic communities can be restored to their original species composition and added to the limited existing areas under protection. Restoration ecology is becoming one of the major growth areas in conservation biology, and has its own professional society, the Society for Ecological Restoration, and journals, *Restoration Ecology* and *Ecological Restoration*. However, conservation biologists in this field must take care to ensure that restoration efforts are legitimate, rather than just a public relations cover for environmentally damaging industrial corporations intent on continuing business as usual (Holloway 1994; Zedler 1996). A 5 ha demonstration project in a highly visible location does not compensate for thousands or tens of thousands of hectares damaged elsewhere, and it should not be accepted as such by conservation biologists. The best long-term strategy is still to protect and manage biological communities where they are naturally found; only in these places can we be sure that the requirements for the long-term survival of all species are available.

Summary

1. Protecting habitat is the most effective method of preserving biological diversity. The area of legally protected habitats worldwide will probably never significantly exceed 10% of the Earth's land surface due to the demands of human societies for natural resources. Well-selected protected areas can initially protect large numbers of species, but their long-term effectiveness remains in doubt.

2. Government agencies and conservation organizations are now setting national and worldwide priorities for establishing new protected areas based on the relative distinctiveness, endangerment, and utility of the species and biological communities that occur in a particular location. To be effective at preserving bio-

logical diversity, Earth's protected areas must include examples of all biological communities.

3. Principles of conservation biology need to be considered along with common sense and experience in designing new protected areas. In general, new parks should be as large as possible and should not be fragmented by roads, fences, and other human constructs. Many endangered species require such undisturbed conditions for their continued existence.

4. Protected areas often must be managed in order to maintain their biological diversity because the original conditions of the area have been altered by human activities. Parts of protected areas may have to be periodically burned, grazed, or otherwise disturbed by people to maintain the habitat types and successional stages that certain species need.

5. Considerable biological diversity exists outside of protected areas, particularly in habitat managed for multiple-use resource extraction. Governments and private landowners are increasingly including the protection of biological diversity over large areas as one of their management priorities for multiple-use land, a practice sometimes called ecosystem management.

6. Restoration ecology provides methods for reestablishing species, whole communities, and ecosystem functions in degraded habitat. Restoration ecology provides an opportunity to enhance biological diversity in habitats that have little other value to humans.

Suggested Readings

Agardy, T. S. 1997. *Marine Protected Areas and Ocean Conservation*. R. G. Landes Company, Austin, Texas. A wide range of planning, design, and policy issues are discussed by a leading authority.

Bennett, A. F. 1999. *Linkages in the Landscape: The Role of Corridors and Connectivity in Wildlife Conservation*. IUCN, Gland, Switzerland. Book-length review of conservation corridors, with theory and examples.

Brandon, K., K. H. Redford and S. E. Sanderson, (eds.). 1998. *Parks in Peril: People, Parks and Protected Areas*. Island Press, Washington, D.C. Managing national parks and providing for the needs of nearby people is a great challenge, and sometimes the problems are overwhelming.

Christensen, N. L. and 12 others. 1996. The report of the Ecological Society of America committee on the scientific basis for ecosystem management. *Ecological Applications* 6: 665–691. Special issue devoted to this topic, with many excellent articles.

Flather, C. H., M. S. Knowles and I. A. Kendall. 1998. Threatened and endangered species geography. *BioScience* 48: 365–376. Endangered species are concentrated in a few hot spot locations.

Forman, R. T. 1995. *Land Mosaics: The Ecology of Landscapes and Regions.* Cambridge University Press, New York. Introductory textbook.

Gilbert, O. L. and P. Anderson. 1998. *Habitat Creation and Repair.* Oxford University Press, Oxford. Practical guide to restoration, with many examples from the United Kingdom.

Karr, J. R. and E. W. Chu. 1998. *Restoring Life in Running Waters.* Island Press, Washington, D.C. Presents a system for monitoring and improving rivers.

Mittermeier, R. A., N. Myers, P. R. Gil and C. G. Mittermeier. 1999. *Hotspots, Earth's Richest and Most Endangered Terrestrial Ecoregions.* Agrupación Sierra Madre, S.C., Mexico City. Lavish, large-format book with pictures and information on each region.

Poiani, K. A., B. D. Richter, M. G. Anderson and H. E. Richter. 2000. Biodiversity conservation at multiple scales: Functional sites, landscapes and networks. *BioScience* 50: 133–146. Protecting biodiversity requires protecting the variation in natural processes, along with ecosystems and species.

Restoration Ecology and *Ecological Restoration.* Check out these journals to see what is really happening in the field. Available from most college and university libraries and from the Society for Ecological Restoration, 1955 W. Grant Road #150, Tucson AZ 85745 U.S.A.; or contact the society at www.ser.org.

Russ, G. R. and A. C. Alcala. 1996. Marine reserves: Rates and patterns of recovery and decline of large predatory fish. *Ecological Applications* 6:947–961. Case study of islands in the Philippines shows increases in fish populations following the establishment of marine reserves. Contains an extensive bibliography on marine reserves.

Shafer, C. L. 1990. *Nature Reserves: Island Theory and Conservation Practice.* Smithsonian Institution Press, Washington, D.C. Comprehensive, well-illustrated review of the theories of reserve design that also presents evidence and counterarguments.

Stattersfield, A. J., M. J. Crosby, A. J. Long and D. C. Wege. 1998. *Endemic Bird Areas of the World: Priorities for Biodiversity Conservation.* BirdLife International, Cambridge. Highlights areas that need additional protection.

Yaffee, S. L. et al. 1996. *Ecosystem Management in the United States: An Assessment of Current Experience.* Island Press, Washington, D.C. Summary descriptions of 105 selected projects, along with listings and contact information for 619 projects. A great source of information.

Zedler, J. B. 1996. Ecological issues in wetland mitigation: An introduction to the forum. *Ecological Applications* 6:33–37. Special issue provides excellent information on the creation of new wetlands.

Chapter *5*

Conservation and Sustainable Development

Many problems in conservation biology require a multidisciplinary approach that addresses the need to protect biological diversity *and* provide for the economic welfare of people. The example of macaw conservation in Chapter 1 illustrates such an approach: Conservation biologists in Peru are training Indian groups at the local level as park guides and naturalists, developing tourist facilities and other business activities to provide economic opportunities for the local people, interacting with the local and national governments to establish protected areas and local land titles, and lobbying at the international level for restricting trade in endangered birds. Conservation biologists throughout the world are actively working to develop such innovative approaches. While some successes can be identified, conservation biology is a young discipline, and there are many challenges in developing general approaches that work in practice.

As we have already seen, efforts to preserve biodiversity sometimes conflict with human needs, both real and perceived

5.1 Sustainable development seeks to address the conflict between development to meet human needs and the preservation of the natural world. (From Gersh and Pickert 1991; drawing by Tamara Sayre.)

(Figure 5.1). Increasingly, many conservation biologists are recognizing the need for **sustainable development**—economic development that satisfies both present and future human requirements for resources and employment while minimizing its impact on biological diversity (Lubchenco et al. 1991).

The concept of sustainable development can be applied in a variety of ways. As defined by some environmental economists, *development* refers to improvements in organization but not necessarily increases in resource consumption, and it is clearly distinguished from *growth*, which denotes material increases in the amounts of resources used. Sustainable development is a useful concept in conservation biology because it emphasizes improving current development and limiting growth. By this definition, then, investing in national park infrastructure to improve the protection of biological diversity and provide revenue opportunities to local people would be an example of sustainable development, as would efforts to develop less destructive logging or fishing practices.

Unfortunately, the concept is often misappropriated: Many large corporations, and the policy organizations that they fund, misuse the concept of sustainable development to "greenwash" their ongoing industrial activities without any change in their practices (Willers 1994). For instance, a plan to establish a huge mining complex in the middle of a forest wilderness cannot justifiably be called "sustainable development" simply because a small percentage of the land area is set aside as a park. Alternatively, some conservationists champion the opposite extreme, claiming that sustainable development means that vast areas of the world be kept off-limits to any development at all and allowed to remain or return to wilderness (Mann and Plummer 1993). As with all controversies, informed scientists and citizens must study the issues carefully, identify which groups are advocating which positions and why, and then make careful decisions that best meet the sometimes contradictory needs of human society and the protection of biological diversity. Such contradiction necessitates compromise, and in most cases, compromises form the basis of government policy and laws, with conflicts resolved by government agencies and courts.

Government Action

Most efforts to find the right balance between the preservation of species and habitats and the needs of society rely on initiatives from concerned citizens, conservation organizations, and government officials, which often end up codified into environmental regulations or laws. These efforts may take many forms, but they begin with individual and group decisions to prevent the destruction of habitats and species in order to preserve something of perceived value.

Local legislation

In modern societies, local (city and town) and regional (county, state, provincial) governments pass laws to provide effective protection for species and habitats and at the same time provide development for the continued needs of society (Press et al. 1996; Buck 1996). Such laws are passed because citizens and political leaders feel that these laws represent the will of the majority and provide long-term benefits to the society. Conservation laws regulate activities that directly affect species and ecosystems. The most prominent of these laws govern when and where hunting can occur, the size and number of animals that can be taken, the types of weapons, traps, and other equipment that can be used, and the species of animals that can be taken. In some settled areas and protected areas, hunting and fishing are banned entirely. Hunting and fishing restrictions are enforced through licensing requirements and patrols by game wardens. Similar laws affect the harvesting of plants, seaweeds, and shellfish. These laws allow resources to be used effectively without destructive overharvesting.

Laws that control the manner in which land is used are another means of protecting biological diversity. These laws include restrictions on the following: amount of use of or access to land, types of land use, and generation of pollution. For example, vehicles and even people on foot may be restricted from habitats and resources that are sensitive to damage, such as bird nesting areas, bogs, sand dunes, wildflower patches, and sources of drinking water. Uncontrolled fires may severely damage habitats, so practices contributing to accidental fires, such as campfires, are often rigidly controlled. Zoning laws sometimes prevent construction in sensitive areas such as barrier beaches and floodplains. Even where development is permitted, building permits increasingly are being reviewed to ensure that damage will not be done to endangered species or to wetlands. For major regional and national projects, such as dams, canals, mining and smelting operations, oil extraction, and highway construction, environmental impact statements must be prepared that describe the damage that the proposed projects could cause. Full consideration of environmental impacts is necessary to prevent inadvertent damage to natural resources.

One of the most powerful strategies for protecting biological diversity at the local level is the designation of biological communities as nature reserves. Governments often set aside public lands for various conservation purposes. In many countries, private conservation organizations are among the leaders in acquiring land for conservation efforts (Dwyer and Hodge 1996). In the United States alone, over 800,000 ha of land are protected at a local level by land trusts, which are private, nonprofit corporations established to protect land and natural resources. At the national level, major organizations such as The Nature Conservancy and the Audubon Society have protected an additional 3 million ha in the United States. In Britain, the National Trust has more than 2 million members, and owns over 200,000 ha of land, much of it farmland, including 26 National Nature Reserves, 466 Sites of Special Interest, 355 properties of Outstanding National Beauty, and 40,000 archeological sites (Figure 5.2). In the Netherlands, about half of the protected areas are privately owned (McNeely et al. 1994). A major focus of many of these private reserve networks is nature conservation, often linked to school programs and environmental education. These organizations are collectively referred to as CARTs—Conservation, Amenity, and Recreation Trusts, a name that reflects their various objectives. Jean Hocker, executive director of the Land Trust Exchange, an association of land trust organizations, explains:

> Different land trusts may save different types of land for different reasons. Some preserve farmland to maintain economic opportunities for local farmers. Some preserve wildlife habitat to ensure the existence of an endangered species. Some protect land in watersheds to improve or maintain water quality. Whether biologic, economic, productive, aesthetic, spiritual, educational, or ethical, the reasons for protecting land are as diverse as the landscape itself. (Elfring 1989)

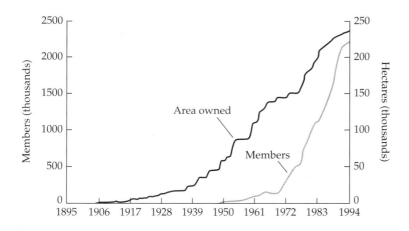

5.2 Membership in the National Trust of Britain has been undergoing a dramatic increase since the 1960s, with a corresponding increase in land ownership. (After Dwyer and Hodge 1996.)

In addition to outright purchases of land, governments, conservation organizations, and private landowners protect land through **conservation easements**, in which landowners give up their rights to develop, build on, or subdivide their property in exchange for a sum of money, a lower real estate tax, or a tax benefit. For many landowners, accepting a conservation easement is an attractive option because they receive a financial advantage while still keeping their land, and because they feel that they are assisting conservation objectives. Of course, the offer of lower taxes or money is not always necessary: Many landowners will voluntarily accept conservation restrictions without compensation. Another option that land trusts use is **limited development**, in which a landowner, a property developer, and a conservation organization reach a compromise that allows part of the land to be commercially developed while the remainder, often the most significant habitat, is protected by a conservation easement. Limited development projects are often successful because the developed lands typically increase in value by being adjacent to conservation land. Limited development also allows the construction of necessary buildings for an expanding human society.

Local efforts by land trusts to protect land are sometimes criticized as elitist because they provide tax breaks to those wealthy enough to take advantage of them, in addition to lowering the revenue collected from land and property taxes. Others argue that land in alternative uses, such as agriculture or shopping malls, is more productive. While land in trust may yield lower tax revenues, the loss of revenue from land acquired by a land trust is often offset by the increased value of property adjacent to the reserve. In addition, the employment, nature tours, tourist spending, and student projects associated with nature reserves, national parks, wildlife refuges, and other protected areas generate revenue throughout the local economy, which benefits the community (Power 1991). Finally, by preserving important features of the landscape and the natural communities, local nature reserves also preserve and enhance the cultural heritage of the local society, considerations that must be included for sustainable development to be achieved.

National legislation

Throughout much of the modern world, national governments play a leading role in conservation activities. The establishment of national parks is a common conservation strategy. National parks are the single largest source of protected lands in many countries. For example, Costa Rica's national parks and protected areas cover 700,000 ha, or almost 14% of the nation's total land area (WRI 1998). Outside of the parks, deforestation is proceeding rapidly, and soon the parks may represent the only undisturbed habitat in the country. As of 1998, the U.S. National Park system, with 379 sites, covered 35 million ha (U.S. National Park Service 2000a,b).

National legislatures and government agencies are the principal instruments for developing national standards on environmental pollution. Laws regulating aerial emissions, sewage treatment, waste dumping, and development of wetlands are often enacted to protect human health and property as well as natural resources such as drinking water, forests, and commercial and sport fisheries. The effectiveness with which these laws are enforced determines a nation's ability to protect the health of its citizens and the integrity of its natural resources. At the same time, these laws protect biological communities that would otherwise be destroyed by pollution: The air pollution that exacerbates human respiratory disease also damages commercial forests and natural biological communities, and the pollution that ruins drinking water also kills terrestrial and aquatic species.

National governments can also have a substantial effect on the protection of biological diversity through control of their borders, ports, and commerce. To protect forests and regulate their use, governments can ban logging, as did Thailand following disastrous flooding; restrict the export of logs, as was done in Indonesia; and penalize timber companies that damage the environment. To prevent the exploitation of rare species, governments can restrict the possession of certain species and control all imports and exports of the species. For example, the export of thick skull bones—a valuable international commodity used for carving—from the rare rhinoceros hornbill bird (*Buceros rhinoceros*) is strictly controlled by the Malaysian government.

Finally, national governments can identify endangered species within their borders and take steps to conserve them, such as acquiring habitat for the species, controlling use of the species, developing a research program that studies the species, and implementing in situ and ex situ recovery plans. In the United States, this is often done through the Endangered Species Act. In European countries, this is accomplished through domestic enforcement of international treaties, such as the Convention on International Trade in Endangered Species and the Ramsar Wetlands Convention.

Traditional Societies and Sustainable Development

As discussed in Chapters 2 and 4, a great deal of biodiversity exists in places where local people have lived for many generations, using the resources of their environment in a sustainable manner. Local people practicing a traditional way of life in rural areas, with relatively little outside influence in terms of modern technology, are variously referred to as tribal people, indigenous people, native people, or traditional people (Dasmann 1991). (These established indigenous people need to be distinguished from more recent settlers, who may not be as concerned with the health of the biological community or as knowledgeable about the species present.) Worldwide, there are approximately 250 million indigenous people in more than 70 countries, occupying 12%–19% of the Earth's total land surface (Redford and Mansour 1996). Indigenous populations who practice their traditional culture are on the decline, however. In most areas of the world, local people are increasingly coming into contact with modernity, resulting in a changing belief system (particularly among the younger members of the society), and a greater use of outside manufactured goods.

People have lived in nearly every terrestrial ecosystem of the world for thousands of years as hunters, fishermen, farmers, and gatherers. Even remote tropical rain forests that are designated as "wilderness" by governments and conservation groups often have a small, sparse human population. In fact, tropical areas of the world have had a particularly long association with human societies because the Tropics have always been free of glaciation and are particularly amenable to human settlement. The great biological diversity of the Tropics has coexisted with human societies for thousands of years, and in most places, humans did not substantially damage the biological diversity of their surroundings (Gomez-Pompa and Kaus 1992). Traditional societies utilizing innovative irrigation methods and a mixture of crops have often been able to support relatively high human population densities without destroying their environment or the surrounding biological communities. The present mixture and relative densities of plants and animals in many biological communities may reflect the historic activities of people in those areas, such as selective hunting of certain game animals, fishing, and planting useful species (Redford

1992). The commonly practiced agricultural system known variously as swidden agriculture, shifting cultivation, and slash-and-burn agriculture has also affected forest structure and species composition by creating a mosaic of forest patches of different ages. In this system, the trees in an area are cut down, the fallen plant material is burned, and crops are planted in the nutrient-rich ash. After one or several harvests, the nutrients are washed out of the soil by the rain; the farmer then abandons the field and cuts down a new patch of forest for planting. This system works well and does not degrade the environment as long as human population density is low and there is abundant forest land.

In addition to coexisting with their environments without destroying them, local people can also manage their environments to maintain biological diversity, as shown by the traditional agroecosystems and forests of the Huastec Indians of northeastern Mexico (Alcorn 1984). In addition to maintaining permanent agricultural fields and practicing swidden agriculture, the Huastec maintain managed forests—known as *te´lom*—on slopes, along watercourses, and in other areas that are either fragile or unsuitable for intensive agriculture (Figure 5.3). These forests contain over 300 species of plants, from which the people obtain food, wood, and other needed

5.3 A Huastec Indian woman at a *te´lom*, an indigenous managed forest in northeastern Mexico. Here she collects sapote fruit (*Manilkara achras*) and cuttings of a frangipani tree (*Plumeria rubra*) for planting. (From Alcorn 1984; photograph by Janis Alcorn.)

products. Species composition in the forests is altered in favor of useful species by planting and periodic selective weeding. These forest resources provide Huastec families with the means to survive the failure of their cultivated crops. Comparable examples of intensively managed village forests exist in traditional societies throughout the world (Oldfield and Alcorn 1991; Nepstad and Schwartzman 1992; Redford and Padoch 1992).

Conservation ethics of traditional societies

The conservation ethics of traditional societies have been viewed from a variety of perspectives by Western civilization. At one extreme, local people are viewed as destroyers of biological diversity who cut down forests and overharvest game. At the other extreme, traditional people are viewed as "noble savages" living in harmony with nature and minimally disturbing the natural environment. An emerging middle view is that traditional societies are highly varied, and there is no one simple description of their relationship to their environment that fits all groups (Alcorn 1993). In addition to the variation among traditional societies, these societies also vary from within: They are changing rapidly as they encounter outside influences, and there are often sharp differences between the older and younger generations.

Many traditional societies do have strong conservation ethics; they are subtler and less clearly stated than Western conservation beliefs, but they do tend to affect people's actions in their day-to-day lives, perhaps more than Western beliefs (Gomez-Pompa and Kaus 1992; Western 1997). One well-documented example is that of the Tukano Indians of northwestern Brazil (Chernela 1987), who live on a diet of root crops and river fish (Figure 5.4). They have strong religious and cultural prohibitions against cutting the forest along the Upper Río Negro, which they recognize as important to the maintenance of fish populations. The Tukano believe that these forests belong to the fish and cannot be cut by people. They have also designated extensive refuges for fish, and permit fishing along less than 40% of the river margin. Anthropologist Janet M. Chernela observes, "As fishermen dependent upon river systems, the Tukano are aware of the relationship between their environment and the life cycles of the fish, particularly the role played by the adjacent forest in providing nutrient sources that maintain vital fisheries."

Beliefs such as these mean that local people sometimes even take the lead in protecting biological diversity from destruction by outside influences. The destruction of communal forests by government-sanctioned logging operations has been a frequent target of protests by traditional people throughout the world (Poffenberger 1990). In India, followers of the Chipko movement hug trees to prevent logging (Gadgil and Guha 1992). In Borneo, the Penan, a small tribe of

5.4 River fish are the main source of protein for the Tukano people of the Amazon basin, who have strong conservation ethics due to their cultural and religious beliefs: They do not cut the forests that surround them because they believe the land belongs to the fish. (Photograph © Paul Patmore.)

hunter-gatherers, have attracted worldwide attention with their blockades of logging roads entering their traditional forests. In Thailand, Buddhist priests are working with villagers to protect communal forests and sacred groves from commercial logging operations (Figure 5.5). As stated by a Tambon leader in Thailand, "This is our community forest that was just put inside the new national park. No one consulted us. We protected this forest before the roads were put in. We set up a roadblock on the new road to stop the illegal logging. We caught the district police chief and arrested him for logging. We warned him not to come again" (Alcorn 1991). Empowering such local people and helping them to obtain legal title to their traditionally owned lands is often an important component of efforts to establish locally managed protected areas in developing countries.

Local people and their governments

In the developing world, local people typically obtain the products they need—including food, fuelwood, and building materials—from their environment. Without these products, some might not be able to survive. Local people often have established local systems of rights to natural resources, which are sometimes recognized by their governments. When such rights are not recognized, however,

5.5 Buddhist priests in Thailand offer prayers and blessings to protect trees from commercial logging operations. (Photograph by Project for Ecological Recovery, Bangkok.)

and when a new national park is created, for example, or when the boundaries of an existing park are rigidly enforced, people may be denied access to a resource that they have traditionally used or even protected. The common practice of disregarding the traditional rights and practices of local people in order to establish new conservation areas has been termed **ecocolonialism** because of its perceived similarity to the historic abuses of native rights by colonial powers of past eras (Cox and Elmqvist 1997). In effect, the creation of a national park under such circumstances often makes local people into poachers, trespassers, and lawbreakers, even though they have not changed their behavior. Most local people will react negatively and even antagonistically when their traditional rights are curtailed (Poffenberger 1996; Homer-Dixon 1999). If they feel that the park and its resources no longer belong to them, but rather to an outside government, they may begin to exploit the resources of the park in a destructive manner. In order to survive, they may violate the park boundaries, which sometimes results in confrontations with park officers. These confrontations have resulted in violence and even deaths.

An extreme example of such a conflict occurred in 1989, when angry members of the Bodo tribe in Assam, India, killed twelve employees of the Manas National Park and opened the area for farming and hunting (McNeely et al. 1990). The Bodo justified their action on the grounds that they were reclaiming their traditional lands, which had been stolen from them by the British and were not returned to them by the modern Indian government. The fact that Manas had been designated a World Heritage Site, containing endangered species such as the Indian rhinoceros (*Rhinoceros unicornis*) and the pygmy hog (*Sus salvanius*), was not relevant to the Bodo; the advantages of the national park were not apparent to them.

In the developing world, it is often not possible to create a rigid separation between lands used by local people to obtain natural resources and strictly protected national parks. Many examples exist in which people are allowed to enter protected areas periodically to obtain natural products. Biosphere Reserves, an international land use designation, allow local people to use resources in designated buffer zones, for instance. Another example is Chitwan National Park in Nepal, which allows local people to collect canes and thatch (Figure 5.6; Lehmkuhl et al. 1988). In addition, large game animals are legally harvested for meat in many African game reserves (Lewis et al. 1990). Agreements have been negotiated between local people and governments allowing cattle to graze inside certain African parks, and in exchange, wild animals outside the parks are unmolested. Through such compromises, the economic needs of local people are included in local conservation management plans, to the benefit of both the people and the reserves. Such compromises, known as **integrated conservation–development projects**, are increasingly being regarded as conservation strategies worthy of serious consideration (Wells and Brandon 1992; Primack et al. 1998; Caldecott 1996; Maser 1997).

Biological diversity and cultural diversity

Biological and cultural diversity are often linked. The rugged tropical areas of the world where the greatest concentrations of species are found are frequently the areas where people have the greatest cultural and linguistic diversity. The geographical isolation that is created by mountain ranges and complex river systems and favors biological speciation also favors the differentiation of human cultures. The cultural diversity found in places such as Central Africa, Amazonia, New Guinea, and Southeast Asia represents one of the most valuable resources of human civilization, providing unique insights into philosophy, religion, music, art, land use, and psychology (Denslow and Padoch 1988). The protection of these traditional cultures within their natural environments provides the opportunity to achieve the dual objectives of protecting biological diversity and

5.6 Local residents collect cane grass and thatching materials from Chitwan National Park in Nepal. Park officials weigh the bundles in order to keep the harvest at a sustainable level. (Photograph © John E. Lehmkuhl.)

preserving cultural diversity. In the words of Victor Toledo, a Mexican conservation biologist, describing his own country (1988):

> In a country that is characterized by the cultural diversity of its rural inhabitants, it is difficult to design a conservation policy without taking into account the cultural dimension; the profound relationship that has existed since time immemorial between *nature* and *culture.* . . . Each species of plant, group of animals, type of soil and landscape nearly always has a corresponding linguistic expression, a category of knowledge, a practical use, a religious meaning, a role in ritual, an individual or collective vitality. To safeguard the natural heritage of the country without safeguarding the cultures which have given it feeling is to reduce nature to something beyond recognition, static, distant, nearly dead.

Cultural diversity is strongly linked to the genetic diversity of crop plants. In mountainous areas in particular, geographically isolated cultures develop local plant varieties known as **land races**; these cultivars are adapted to the local climate, soils, and pests, and they satisfy the tastes of the local people. The genetic variation in these land races has global significance to modern agriculture because of its potential to improve crop species (see Chapter 3).

Conservation efforts involving traditional societies

Several strategies exist for integrating the protection of biological diversity, the customs of traditional societies, and the genetic variability of traditional crops. Many of them could be classified as integrated conservation–development projects. A large number of such programs have been initiated over the last decade, leading to opportunities for evaluation and improvement (Salafsky and Margoluis 1999).

Biosphere reserves. UNESCO's Man and the Biosphere Program (MAB) includes among its goals the maintenance of "samples of varied and harmonious landscapes resulting from long-established land use patterns" (Batisse 1997). The MAB Program recognizes the role of people in shaping the natural landscape, as well as the need to find ways in which people can sustainably use natural resources without degrading the environment. The MAB research framework, applied in its worldwide network of designated Biosphere Reserves (see Chapter 4), integrates natural science and social science research. It includes investigations of how biological communities respond to different human activities, how humans respond to changes in their natural environment, and how degraded ecosystems can be restored to their former conditions.

One valuable example of a Biosphere Reserve is the Kuna Yala Indigenous Reserve on the northeastern coast of Panama (Gregg 1991). In this protected area of 60,000 ha of tropical forest, 30,000 Kuna people in 60 villages practice traditional medicine, agriculture, and forestry, while documentation and research are conducted by scientists from outside institutions. The Kuna carefully regulate the levels of scientific research in the reserve: they insist on local training, require scientists to present reports before they leave the area, charge research fees, and oblige the scientists to be accompanied by local guides. The Kuna people even control the type and rate of economic development in the reserve, with assistance from their own outside, paid advisors. The level of empowerment of the Kuna people is unusual and illustrates the potential for a traditional society to take control of its destiny, culture, and environment. Unfortunately, empowering traditional people is no guarantee that biodiversity will be preserved; particularly as traditions change or disappear, economic pressures for exploitation increase, and programs are mismanaged (Oates 1999). Such a change appears to be occurring in the Kuna: As traditional conservation beliefs erode in the face of outside influences, younger Kuna are beginning to question the need to rigidly protect the reserve (Redford and Mansour 1996).

In situ agricultural conservation. In many areas of the world, local farmers cultivating locally adapted varieties of crop plants can preserve genetic variability in crop species (Figure 5.7). For example, thousands of distinct varieties of potatoes are grown by Andean farmers in South America. Often these farmers grow many varieties in one field to minimize the risk of crop failure and to produce different varieties for different uses. Similarly, traditional farmers in the Apo Kayan of Borneo may grow more than 50 varieties of rice. These local varieties often have unique genes for dealing with disease, nutrient deficiencies, pests, drought, and other environmental variations (Cleveland et al. 1994). Moreover, these local varieties continue to evolve new genetic combinations, some of which may be effective in dealing with looming global environmental threats. However, farmers throughout the world are abandoning their traditional forms of agriculture with local land races in order to grow high-yielding varieties using capital-intensive methods. In countries such as Indonesia, Sri Lanka, and the Philippines, over 80% of the farmers have adopted modern varieties.

While an increased yield may be better in the short term for the farming families and their countries, the long-term health of modern agriculture depends on the preservation of the genetic variabil-

5.7 It is useful to view traditional agricultural practices both from a human cultural and an agricultural perspective. A synthesis of these viewpoints can lead to theoretical and methodological approaches toward the conservation of the environment, the culture, and the genetic variation found in these traditional agroecosystems. (After Altieri and Anderson 1992.)

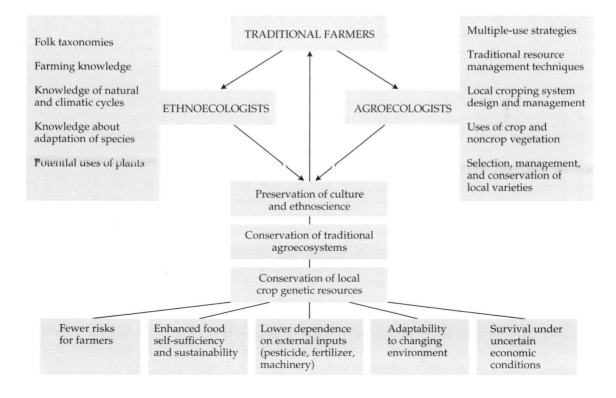

ity represented by local varieties. One innovative suggestion has been for an international agricultural body to subsidize villages to be in situ (in place) "land race custodians" (Wilkes 1991; Altieri and Anderson 1992). Villagers in these zones would be paid to grow their traditional crops in a traditional manner, providing a crucial source of genes for modern crop improvement programs.

Programs incorporating in situ conservation practices have already been initiated in some places. In Mexico in particular, a number of development programs are attempting to integrate traditional agriculture, conservation, and research (Benz et al. 1990; Toledo 1991). One example is the 140,000 ha Sierra de Manantlán Biosphere Reserve in western Mexico, which was established to preserve the only known populations of *Zea diploperennis,* a perennial relative of corn (maize); additionally, the reserve protects a rich subtropical forest with numerous species new to science. This plant, occurring only in abandoned *milpas* (fields planted using traditional shifting agricultural methods), is of great potential value in efforts to preserve genes that may someday be used to protect the annual corn crop, valued at $55 billion per year (Figure 5.8). Imagine a disease-resistant perennial maize crop that did not have to be planted each year. In this instance, the long-term protection of *Z. diploperennis* in the wild depends on encouraging local farmers to remain on the land and continue their traditional cultivation practices.

A slightly different approach used in arid regions of the American Southwest involves linking traditional agriculture and genetic conservation (Nabhan 1989). A private organization called Native Seeds/SEARCH collects the seeds of traditional crop cultivars for long-term preservation. The organization also encourages farmers to grow traditional crops, provides them with the seeds of traditional cultivars, and buys the farmers' unsold production. Countries have also established special reserves to conserve areas containing wild relatives of crops. For example, species reserves protect the wild relatives of wheat, oats, and barley in Israel and citrus in India.

Extractive reserves. In many areas of the world, indigenous people have extracted natural products from their surrounding biological communities for decades and even centuries. The sale and barter of these products constitute a major part of the people's livelihoods. Understandably, local people are very concerned about retaining their rights to continue collecting natural products. In areas where such collection represents an integral part of the indigenous society, the establishment of a national park that excludes traditional collecting will meet with as much resistance from the local community as will a land-grab by outsiders that involves exploitation of natural resources and a conversion of the land to other uses. A new type of

5.8 *Zea diploperennis* is a perennial relative of maize grown in the Sierra de Manantlán Biosphere Reserve in Mexico. American scientist Mike Nee is shown here with a bundle of harvested *Zea diploperennis*. (Photograph by Hugh Iltis.)

protected area known as an **extractive reserve** may present a sustainable solution to this problem.

The Brazilian government is trying to meet the legitimate demands of local citizens through extractive reserves, from which local people collect natural products such as rubber, resins, and Brazil nuts, in a way that minimizes damage to the forest ecosystem (Murrieta and Rueda 1995). Such areas, currently comprising about 3 million hectares, guarantee the ability of local people to continue their way of life and guard against possible conversion of the land to cattle ranching and farming. This land use also serves to protect the biological diversity of the area because the ecosystem remains basically intact. Despite these advantages, however, populations of large animals in extractive reserves may still decline due to subsistence hunting.

The real challenge for local people who rely on extractive reserves to collect their natural products is to develop products that can be sold at a good market price. If the local people cannot survive by collecting and selling natural products, they might be forced, out of economic desperation, to cut down their forests for timber and use the land for agriculture. Because many extractive reserves are only able to operate with the support of outside subsidies, the economic viability of these reserves is still unknown.

Community-based initiatives. In many cases, local people already protect the forests, rivers, coastal waters, wildlife, and plants in the vicinity of their homes. Such protection is often enforced by village elders and is based on religious and traditional beliefs. Governments and conservation organizations can assist these local conservation initiatives by providing help in obtaining legal title to traditional lands, access to scientific expertise, and financial assistance to develop needed infrastructure. One example of such a project is the Community Baboon Sanctuary in eastern Belize, created by a collective agreement among a group of villages to maintain the forest habitat required by the local population of howler monkeys (*Alouatta pigra*) (Horwich and Lyon 1998). Ecotourists visiting the sanctuary must pay a fee to the village organization and additional fees if they stay overnight and eat meals with a local family. Conservation biologists working at the site have provided training for local nature guides, a body of scientific information on the wildlife, funds for a local natural history museum, and business training for the village leaders.

In the Pacific islands of Samoa, much of the rain forest land is under "customary ownership"—it is owned collectively by communities of indigenous people (Cox 1997). Villagers are under increasing pressure to sell logs from their forests to pay for schools and other necessities. Despite this situation, the local people have a strong desire to preserve the forests because of their religious and cultural significance, as well as their value as sources of medicinal plants and other products. A variety of solutions are being developed to meet these conflicting needs. In American Samoa, the U.S. government leased forest and coastal land from the villages to establish a new national park. In this case, the villages retained ownership of the land and traditional hunting and collecting rights. Village elders were also assigned places on the park advisory board. In Western Samoa, international conservation organizations and various donors built schools, medical clinics, and other public works projects that the villages needed in exchange for their stopping all commercial logging. Thus, each dollar donated did double service, both protecting the forest and providing humanitarian aid to the villages.

A key element in the success of these community-based projects was the ability to build on and work with stable, flexible local institutions. Conservation initiatives involving recent immigrants or demoralized local people are generally more difficult. In addition, while working with local people may be a worthy ultimate goal, in some cases it may not be possible. Sometimes the only way to preserve biological diversity will be to exclude people from protected areas and rigorously patrol their boundaries. A key point to note is the fact that regardless of whether they support or oppose conservation activities, people are more likely to respond to issues that affect their everyday lives. In many cases, improving the economic conditions of people's lives will need to be part of the strategy of preserving biological diversity in developing countries (Hackel 1999). The challenge for conservation biologists is to energize local people in support of conservation while addressing the objections of those who oppose it.

International Approaches to Conservation and Sustainable Development

Biological diversity is concentrated in the tropical countries of the developing world, most of which are relatively poor and experiencing rapid rates of population growth, development, and habitat destruction. Developing countries are often willing to preserve biological diversity, but they may be unable to pay for the habitat preservation, research, and management required for the task. The developed countries of the world (including the United States, Canada, Japan, Australia, and many of the European nations) require the biological diversity of the Tropics to supply genetic material and natural products for agriculture, medicine, and industry. How can countries work together to preserve biological diversity?

The Earth Summit

One hallmark of international progress to protect biological diversity is the **Earth Summit**, which was held for 12 days in June 1992 in Rio de Janeiro, Brazil. Known officially as the United Nations Conference on Environment and Development (UNCED), in attendance were representatives from 178 countries, over 100 heads of state, plus leaders of the United Nations and major nongovernmental and conservation organizations. The purpose of the conference and associated follow-up conferences was to discuss ways of combining increased protection of the environment with more effective economic development in less wealthy countries (United Nations 1993a,b) (Figure 5.9). The conference was successful in heightening awareness of the seriousness of the environmental crisis and placing the issue at the center of world attention (Haas et al. 1992). A noteworthy feature of the conference was that attendants seemed to perceive a clear connection between the protection of the environ-

5.9 World leaders met in Rio de Janiero in 1997 for "Rio +5," a follow-up to the 1992 Earth Summit. Mikhail Gorbachev, the former president of the Soviet Union, is shown here addressing a forum of prominent political and environmental leaders. (Photograph by Hiromi Kobori.)

ment and the need to alleviate the poverty of the developing world through increased levels of financial assistance from wealthier countries. The conference participants discussed and eventually signed the major documents described below:

- *The Rio Declaration*. The right of nations to utilize their own resources for economic and social development is recognized, as long as the environment elsewhere is not harmed. The Declaration affirms the "polluter pays" principle: companies and governments take financial responsibility for the environmental damage that they cause.

- *Convention on Climate Change*. This agreement and subsequent follow-up agreements require industrialized countries to reduce their emissions of carbon dioxide and other greenhouse gases to 1990 levels, and to make regular reports on their progress. Developing countries must report their levels of emission.

- *Convention on Biodiversity*. The **Convention on Biological Diversity** has three objectives: the protection of biological diversity; its sustainable use; and equitable sharing of the benefits of new products made with wild and domestic species. While the first two objectives are straightforward, the last addresses a thorny issue by recognizing that developing countries should receive fair compensation for the use made of species collected within their borders. This treaty has been ratified by almost 170 nations so far, but the United States Congress has delayed approving the convention because of what it perceives as restrictions on the enormous U.S. biotechnology industry. Developing specific agreements among countries and biotechnology companies to implement sharing of profits from new products has often proved to be difficult and complex. In many cases, tropical developing countries have severely restricted or even halted the removal of any biological materials from their countries. This has had the unexpected effect of hindering international scientific research unrelated to biological prospecting.

- *Agenda 21.* This 800-page document is an innovative attempt to describe in a comprehensive manner the policies needed for environmentally sound development. Agenda 21 shows the links between the environment and other issues that are often considered separately, such as child welfare, poverty, women's issues, technology transfer, and unequal divisions of wealth. Plans of action are described to address problems of the atmosphere, land degradation and desertification, mountain development, agriculture and rural development, deforestation, aquatic environments, and pollution. This document also describes financial, institutional, technological, and legal mechanisms for implementing these action plans.

The most contentious issue at the conference was deciding how to fund the Earth Summit programs, particularly Agenda 21. The cost of these programs was estimated to be about $600 billion per year, of which $125 billion was to come from the developed countries as overseas development assistance (ODA). Because the existing levels of ODA at that time amounted to $60 billion per year for all activities, this meant that implementing Agenda 21 would have required a tripling of the yearly foreign aid commitment. The major developed countries, known as the Group of Seven, did not agree to this increase in funding, but did agree in principal to raise their levels of foreign assistance to 0.7% of gross national product by the year 2000. Only a few wealthy northern European countries, most notably Norway, Denmark, Sweden, the Netherlands, and Finland, have met this target, while the larger developed countries, including the United States, have actually lowered their funding levels.

In the end, however, a major new source of funds for conservation and environmental activities in developing countries was created at the time of the Earth Summit. The **Global Environment Facility** (GEF) was established by the World Bank and the developed countries, along with the United Nations Development Programme (UNDP) and the U.N. Environmental Programme (UNEP). The GEF was established as a pilot program with current financing of $2.75 billion, to be used for funding projects relating to global warming, biodiversity, international waters, and ozone depletion. Two evaluations of the GEF judged the first round of projects to be a mixed success, and identified lack of participation by community groups and government leaders as a major problem (Bowles and Prickett 1994; UNDP/UNEP/World Bank 1994). An additional problem was the mismatch of large-scale funding over short periods with the long-term needs of poor countries. Many projects failed to deal with the "3 Cs" of community development: genuine *concern* from all parties to solve the problem, enforceable *contracts* to ensure the work is done satisfactorily, and *capacity*—that is, the trained people, equipment, and facilities to carry out the work. One of the most important results of the Earth Summit is the series of meetings that continue to be held to carry out the recommendations. The most notable success has been the international agreement, reached at Kyoto in December of 1997, to reduce industrialized countries' global greenhouse gas emissions to below 1990 levels.

Funding sustainable development programs

How are funds allocated to conservation and development projects? When a conservation need, such as protecting a species or establishing a nature reserve, is identified, it often initiates a complex process of project design, proposal writing, fund-raising, and implementation that involves several different types of conservation organizations. Charitable foundations (e.g., the MacArthur Foundation), government agencies (e.g., the U.S. Agency for International Development), and international institutions (e.g., the World Bank) often provide money for conservation projects through direct grants to the institutions that implement the projects (e.g., Colorado State University, the Missouri Botanical Garden, the Wildlife Conservation Society). In other cases, foundations and government agencies give money to major nongovernmental conservation organizations (e.g., World Wildlife Fund, Conservation International), which in turn provide grants to local conservation organizations. The major international conservation organizations are often active in establishing, strengthening, and funding local nongovernmental organizations, as well as government conservation programs, in the developing world. Working with local organizations in developing countries is an effective strategy because it provides training and support for groups of citizens within the country, who can then be advocates for conservation for many years to come.

International funding

Increasingly, groups in the developed countries are realizing that if they want to preserve biological diversity in species-rich but cash-poor developing countries, they cannot simply provide advice: a financial commitment is also required. Institutions within the United States, Japan, Germany, Britain, and other developed countries represent some of the largest sources of this financial assistance. The aid these organizations provide for conservation and sustainable development is substantial: In 1991, a total of 1410 projects receiving aid from U.S. institutions were identified in 102 developing countries, accounting for a total investment of $105 million (Abramovitz 1991, 1994). The predominant sources of funds were U.S. government agencies ($70 million) such as the Agency for International Development and the National Science Foundation; charitable foundations ($20 million) such as the Mellon Foundation, the MacArthur Foundation, the W. Alton Jones Foundation, Pew Charitable Trusts; and nongovernmental organizations ($10 million) such as the World Wildlife Fund, Conservation International, and The Nature Conservancy. In the United States, investment by large foundations increased sevenfold between 1987 and 1991, and government funding tripled during this period, demonstrating that conservation and sustainable development have clearly been targeted as funding priorities (Figure 5.10).

While funding levels for conservation in developing countries are increasing substantially, the amount of money being spent is still inadequate to protect the great storehouse of biological riches needed for the long-term prosperity of human society. Compared with the $13 billion spent yearly on the U.S. space program and the $315 million spent each year on the Human Genome Project, the $105 million per year being spent by United States organizations to protect biological diversity in developing countries is meager indeed.

Funding in developing countries

In addition to direct grants for projects, an increasingly important mechanism used to provide secure, long-term support for conservation activities in developing countries is the **national environmental fund** (**NEF**). NEFs are typically set up as conservation trust funds or foundations in which a board of trustees—composed of representatives of the host government, conservation organizations, and donor agencies—allocates the annual income from an endowment to support inadequately funded government departments, as well as nongovernmental conservation organizations and activities. NEFs have been established in over 20 countries with funds contributed by the governments of developed countries and by such major organizations as the World Bank and the World Wildlife Fund (IUCN/TNC/WWF 1994; Mitikin and Osgood 1994).

One important early example of an NEF, the Bhutan Trust Fund for Environmental Conservation (BTF), was established in 1991 by

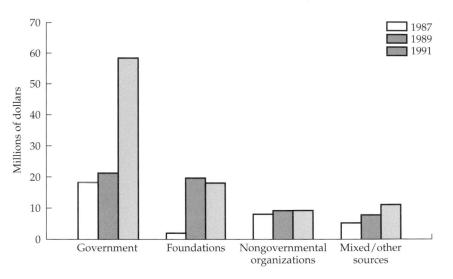

5.10 Funding from U.S. sources for biodiversity research and conservation efforts in developing countries, by type of funding organization: organizations and divisions of the U.S. government; charitable foundations and trusts; nongovernmental organizations such as the World Wildlife Fund; and funding from multiple sources or from miscellaneous institutions (such as universities, zoos, and museums). (From Abramovitz 1994.)

the government of Bhutan in cooperation with the United Nations Development Programme and the World Wildlife Fund. The BTF has already received about $14 million out of its goal of $20 million, with the Global Environment Facility being the largest donor. Activities include surveying the rich biological resources of this eastern Himalayan country, training foresters, ecologists, and other environmental professionals, promoting environmental education, establishing and managing protected areas, and designing and implementing integrated conservation development projects.

Another innovative idea, the **debt-for-nature swap**, has used international debt as a vehicle for financing projects to protect biological diversity. Collectively, developing countries owe about $1.3 trillion—which represents 44% of their combined gross national products—to international financial institutions (Dogsé and von Droste 1990). The commercial banks that hold these debts sometimes sell them at a steep discount on the international secondary debt market, due to low expectations of repayment. Thus, an international conservation organization is able to buy a developing country's discounted debt from a bank. The debt is then canceled in exchange for the developing country's agreement to make annual payments, in its own currency, for conservation activities such as land acquisition, park management, and public education.

In other "swaps," governments of developed countries owed money directly by developing countries may decide to cancel a certain percentage of the debt if the developing country will agree to contribute to a national environmental fund or other conservation activity. Debt-for-nature swaps have converted debt valued at over $1.5 billion for conservation and sustainable development activities in Costa Rica, Colombia, Poland, Madagascar, and a dozen other countries. In a recent program, Spain agreed to allow Latin American countries to convert $70 million in debt to conservation programs of equivalent value, and many European countries are planning to extend such programs, particularly to Poland.

The total amount of debt involved in debt-for-nature swaps is only about 0.1% of the total debt owed by the developing world, so the overall effect on the economies of individual countries has been minimal so far. Debt swaps have also proved to be complex and difficult to negotiate due to their novelty, the weak financial condition of many debtor governments, and shifting political climates. Although their application appears to be fairly limited, in particular situations, debt-for-nature swaps can be one of many useful financing tools (Thapa 1998).

International development banks and ecosystem damage

The rates of tropical deforestation, habitat destruction, and loss of aquatic ecosystems have sometimes been accelerated by poorly conceived, large-scale projects—sometimes involving dams, roads, mines, and resettlement plans—financed by the international development agencies of major industrial nations, as well as by the four major **multilateral development banks** (**MDBs**) controlled by those nations. These MDBs include the **World Bank**, which lends to all regions of the globe, and the regional MDBs, the Inter-American Development Bank (IDB), the Asian Development Bank (ADB), and the African Development Bank (AFDB).

The MDBs loan more than $25 billion per year to 151 countries to finance economic development projects (Rich 1990). Related to the MDBs are international financial institutions, such as the International Monetary Fund (IMF) and the International Finance Corporation, and government-supported export credit agencies, such as the U.S. Export–Import Bank, Japan's Export–Import Bank, Germany's Hermes Guarantee, Britain's Export Credits and Guarantee Department, France's COFACE, and Italy's SACE: these international institutions collectively support $400 billion of foreign investments and exports each year, almost 8% of total world trade (Kapur et al. 1997; Rich 2000). These export credit agencies exist primarily to support the corporations of developed countries in selling manufactured goods and services to developing countries. While the goal of these financial institutions ostensibly is economic development, the effect of many of the projects they support is to promote growth and exploit natural resources for export to international markets. In many cases, these development projects have resulted in the destruction of ecosystems over a wide area. A study by the World Bank of its own projects found that 37% of the projects were unsatisfactory based on environmental criteria (WRI 1994).

Development lending gone awry: Case studies

Among the most highly publicized examples of environmental destruction resulting from MDB and World Bank lending are the transmigration programs in Indonesia, road construction and development in Brazil, and large dams in places such as Indonesia, India, China, Nepal, and Pakistan.

Indonesian resettlement. From the 1970s to the late 1980s, the World Bank loaned $560 million to the Indonesian government to resettle millions of people from the densely populated inner islands of Java, Bali, and Lombok on the sparsely inhabited, heavily forested outer islands of Borneo (Kalimantan), New Guinea (Irian Jaya), and Sulawesi (Rich 1990). These settlers were supposed to raise crops to feed themselves as well as cash crops, such as rubber, oil palm, and cacao, that could be exported to pay off the World Bank loan. This transmigration program has been an environmental and economic failure because the poor tropical forest soils on the outer islands are not suited to the intensive agriculture practiced by the settlers (Whitten 1987). As a result, many of the settlers have become impoverished and are forced to cut down forests to practice shifting cultivation. The production of export crops to pay off the World Bank loan has not materialized. In addition, at least 2 million and possibly up to 6 million ha of tropical rain forest, as well as adjacent aquatic ecosystems, have been destroyed by the settlers. Although this amount of land is enormous, it still represents only a small fraction of the forested area of the outer islands.

Brazilian highways. The highway projects in the western Brazilian state of Rondônia provide a classic example of a misguided development program (Fearnside 1990; Anderson 1990; Kapur et al. 1997). The World Bank and the Inter-American Development Bank have loaned hundreds of millions of dollars to Brazil since 1981 to build roads and settlement areas in this region. When a highway to Porto Velho, the capital of Rondônia, was opened, farmers from southern and northeastern Brazil, who had been displaced from their land by increasing mechanization and land ownership laws that favored the wealthy, flocked to Rondônia seeking free land. Much of the land in Rondônia was unsuitable for agriculture but was cleared to establish land claims; this practice was often facilitated by tax subsidies. As a result, Rondônia had one of the most rapid rates of deforestation in the world during the 1980s. At the peak of deforestation in 1987, 20 million hectares—2.5% of Brazil's total land area—was burned in one of the world's most massive episodes of environmental devastation. Agricultural, industrial, and transportation projects were consistently launched in this area without environmental impact studies or land use studies to determine their feasibility. In its haste to develop the region, the Brazilian government also built roads across Amerindian reserves and biological reserves that were supposed to be completely protected, effectively opening up even these areas to deforestation. The result was environmental devastation with minor, fleeting economic benefits. Massive forest destruction in Rondônia and elsewhere in Brazil has continued through the 1990s, with particularly high rates of clearing and forest fires in 1997 (Nepstad et al. 1999).

Dam projects. Dams and irrigation systems that provide water for agricultural activities and generate hydroelectric power constitute a major class of projects financed by development agencies and banks (Figure 5.11; Goodland 1990). These projects often damage large aquatic ecosystems by changing water depth and current patterns, increasing sedimentation, and creating barriers to dispersal. As a result of these changes, many species are no longer able to survive in the altered environment. The Three Gorges Dam currently being built on the Yangtze River in China, funded in part by German, Swiss, and

5.11 A hydroelectric dam on the Volta River in Ghana. The watersheds around such dams must be protected if the dams are to operate efficiently. (Photograph courtesy of FAO.)

Canadian export credit agencies, is an example of such a massive project. The environmental impacts of this project will be extensive: It will displace 1.2 to 1.9 million people, flood 250,000 ha of farmland, and cover 8000 sites of cultural significance (Chau 1995).

Ironically, research indicates that the long-term success of some of the large international dam projects that threaten aquatic ecosystems may depend on preserving the forest ecosystem surrounding the project site. The loss of plant cover on the slopes above water projects often results in soil erosion and siltation, with resulting losses of efficiency, higher maintenance costs, and damage to irrigation systems and dams. Protecting the forests and other natural vegetation in the watersheds is now widely recognized as an important and relatively inexpensive way to ensure the efficiency and longevity of water projects, while at the same time preserving large areas of natural habitat. In one study of irrigation projects in Indonesia, it was found that the cost of protecting watersheds ranged from only 1%–10% of the total cost of the project, in contrast to an estimated 30%–40% drop in efficiency due to siltation that would result if the forests were not protected (MacKinnon 1983). One of the most successful examples of an effective environmental investment was the $1.2 million loan made by the World Bank to assist in the development and protection of Dumoga–Bone National Park in northern Sulawesi, Indonesia (McNeely et al. 1990). A 278,700 ha primary rain forest, which included the catchment area on the slopes above a $60 million irrigation project, also financed by the World Bank, was converted into a national park (Figure 5.12). In this particular case, the World Bank was able to protect its original investment through environmental funding representing 2% of the project's cost, and create a significant new national park in the process.

Reforming development lending

Despite repeated criticisms, the MDBs and export credit agencies continue to fund mammoth, environmentally damaging projects in the developing world. Making large loans is the reason that banks exist, and governments want big loans to maintain economic prosperity. What can be done so that multilateral development banks operate more responsibly? First, they can stop making loans for environmentally destructive projects (Kapur et al. 1997). This step would require the banks to analyze development projects using economic cost–benefit models that include the full range of environmental, ecological, and social impacts of projects. Programs that encourage land reform, reductions in rural poverty, establishment of new protected areas, and truly sustainable development should be encouraged. Also, banks need to encourage open public discussion among all groups in a country before projects are implemented.

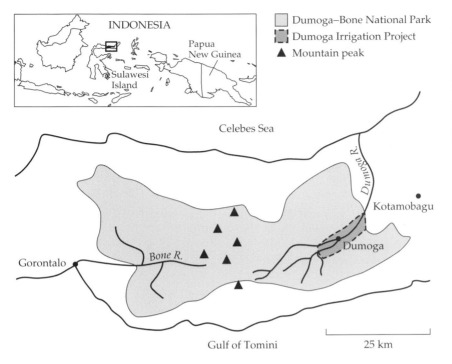

5.12 The Dumoga–Bone National Park on the northern arm of Sulawesi Island, Indonesia, protects the watersheds above the Bone River and the Dumoga River, including the Dumoga Irrigation Project. (After Wells and Brandon 1992.)

Banks are moving in this direction by allowing public examination, independent evaluations, and discussions of environmental impact reports by local organizations affected by planned projects.

Because the MDBs are controlled by the governments of the major developed countries, such as the United States, Japan, Germany, the United Kingdom, and France, their policies are open to scrutiny from the elected representatives of the MDB member countries, their national media, and conservation organizations. In particular, as some of the ill-conceived projects of the World Bank have been publicly criticized, the World Bank has reacted by making the conservation of biological diversity part of its assistance policy and requiring new projects to be more environmentally responsible (Steer 1996). This "new environmentalism" policy recognizes that there is a high cost for inappropriate economic policies, that reducing poverty is often crucial in protecting the environment, and that economic growth must incorporate environmental values. Environmental projects in 52 countries have now been funded by the World Bank, and the total number of projects being funded is increasing (Figure 5.13). However, it remains to be seen whether the MDBs will actually change their basic practices or change only their rhetoric and publicize a few "showcase" projects. It should be noted

5.13 Values of environmental projects that have been funded by the World Bank since 1986. Numbers of projects being funded in a given year are given above the bars. (After Steer 1996).

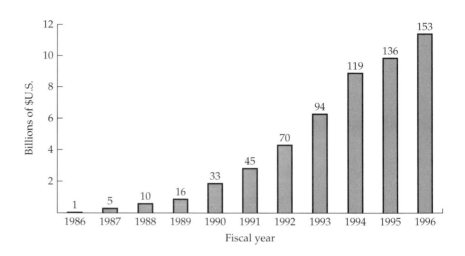

that the MDBs have no enforcement authority; once the money is handed over, countries can choose to ignore the environmental provisions in the agreement, despite local and international protests. However, recent years have seen at least a few examples of loans being held up until governments showed a willingness to address significant environmental issues. Unfortunately, another trend is for the World Bank and the MDBs to finance "clean" projects that can be publicly justified on environmental and social grounds, while the far larger import–export banks quietly support the huge projects that damage the environment and benefit large corporations.

An Agenda for the Future

People at all levels of society must learn that it is in their own interests to work for conservation and to halt the ongoing worldwide loss of species and biological communities. If conservationists can demonstrate that the protection of biological diversity has more value than its destruction, then people and their governments may be willing to take positive action.

There is a consensus among conservation biologists that a number of major problems are involved in preserving biological diversity, and that certain changes in policies and practices are needed. These problems, and suggested responses to them, are listed below.

Problem: Protecting biological diversity is difficult when most of the world's species remain undescribed by scientists, and entire biological communities remain undiscovered.

Response: More scientists and enthusiastic nonscientists need to be trained to identify, classify, and monitor species and biological communities, and funding should be increased in this area, particularly for training scientists and establishing research facilities in the developing world. Natural history societies and nature clubs involving scientists, the general public, and students should be established and encouraged. Information on biological diversity needs to be more accessible; this may be accomplished in part through the new Global Biodiversity Information Facility, which will serve as a central clearinghouse for data from 29 industrialized countries (Redfearn 1999).

Problem: Many conservation issues are global in scope, involving many countries.

Response: Countries are increasingly willing to discuss international conservation issues, as shown by the 1992 Earth Summit, and the 1997 Climate Change Conference in Japan, and to sign and implement treaties such as the recent conventions on biodiversity and global climate change, and trade in endangered species. International conservation efforts are expanding, and further participation in these activities by conservation biologists and the general public should be encouraged.

Problem: Destruction of biological diversity occurs because of overconsumption of the world's resources by the people, business activities, and institutions of developed countries, and the pollution that results from this overconsumption (Figure 5.14).

Response: Citizens, corporations, and countries must take responsibility for their overuse of natural resources (Salonius 1999). Widespread changes in lifestyle, reduced use of resources, and alternative markets for environmentally friendly and recycled products can all have a positive effect on the environment. Specific actions to help conserve natural resources include building more fuel-efficient vehicles, mass transportation systems, and residences, and encouraging their use through incentives and regulation.

Problem: Developed countries often place a greater emphasis on the preservation of biological diversity than do the poorer developing countries that have the most biological diversity.

Response: Developed countries and international conservation organizations should provide secure, long-term financial support to developing countries that establish and maintain national parks and other protected areas. Related economic and social problems must be resolved at the same time, particularly those related to poverty, education, and civil rights. As mandated by the Convention on Biological Diversity, the economic benefits need to be shared in a mutually agreeable manner when new products are developed from wild species collected in the developing country.

Problem: Economic analyses often paint a falsely encouraging picture of development projects that are environmentally damaging.

5.14 In one year, an average U.S. family of four with a typical American lifestyle consumes approximately 3785 liters (1000 gallons) of oil, shown here in barrels, for fueling its two cars and heating its home. The same family burns about 2270 kg (5000 pounds) of coal, shown here in a pile in the right foreground, to generate the electricity to power lights, refrigerators, air conditioners, and other home appliances. The air and water pollution that results from this consumption of resources directly harms biological diversity, and such consumption must be addressed in a comprehensive conservation strategy. (Photograph by Robert Schoen/ Northeast Sustainable Energy Association.)

Response: New types of cost–benefit analyses that include environmental and human costs must be openly discussed, developed, and used; these must factor in costs such as soil erosion, water pollution, loss of natural products, loss of traditional knowledge with potential economic value, loss of tourism potential, loss of species of possible future value, and loss of places to live. Environmental impact analyses also need to include comparative studies of similar projects completed elsewhere, and the probabilities and costs of possible worst-case scenarios.

Problem: Ecosystem services do not receive the recognition they deserve in economic decision making.

Response: Economic activities should be linked with the maintenance of ecosystem services through fees, penalties, and land acquisition. The "polluter pays" principle must be adopted, in which industries, governments, and individual citizens pay for cleaning up the environmental damage their activities have caused (Bernow et al. 1998). A step in this direction is the increasing rates charged to customers by utilities for water and sewer use, rates that reflect the actual costs of providing these services. Financial subsidies to industries that damage the environment and human health—such as the pesticide, petrochemical, logging, fishing, and tobacco industries—should end, and those funds shifted to activities that enhance the environment and human well-being.

Problem: Much of the destruction of the world's biological diversity is caused by people who are desperately poor and are simply trying to survive.

Response: Conservation biologists and charitable and humanitarian organizations need to assist local people in organizing and developing suitable economic activities that do not damage biological diversity. In many cases, this will involve using existing cropland more efficiently rather than increasing the area of cultivation and human impact (Goklany 1998). Foreign assistance programs need to be carefully planned so that they help to alleviate rural poverty rather than primarily benefiting urban elites, as they often do. Programs promoting smaller families and reducing human population growth should be closely linked to efforts aimed at improving economic opportunities and halting environmental degradation (Dasgupta 1995).

Problem: Decisions on land conversion and the establishment of protected areas are often made by central governments with little input from people in the region being affected. Consequently, local people sometimes feel alienated from conservation projects and do not support them.

Response: Local people have to believe that they will benefit from a conservation project and that their involvement is important. To achieve this goal, environmental impact statements and other project information should be publicly available to encourage open discussion at all steps of a project. Decision-making mechanisms should be established to ensure that the rights and responsibilities of management of conservation projects are shared between government agencies and local communities.

Problem: Revenues, business activities, and scientific research associated with national parks and other protected areas often do not directly benefit surrounding communities.

Response: Whenever possible, local people should be trained and employed in parks as a way of utilizing local knowledge and providing local income. Also, a portion of park revenues can be used to fund community projects such as schools, clinics, and community-run businesses, which directly benefit people living near the park. Conservation biologists working in national parks should periodically explain the purpose and results of their work to nearby communities and to school groups, and they should listen to what the local people have to say.

Problem: National parks and conservation areas in developing countries have inadequate budgets to pay for conservation activities. Revenues that they collect are often returned to government treasuries.

Response: Increased funds for park management can often be raised from foreign tourists and scientists by charging them higher rates for admission, lodging, or meals, which more closely reflect actual maintenance costs. Making sure that these revenues and profits remain at the park and in the surrounding area is important. Also, zoos and conservation organizations in the developed world can make direct financial contributions to conservation efforts in developing countries.

Problem: People cut down tropical forests and plant crops to establish title to the land, even on lands that are not suitable for agriculture. Timber companies that lease forests and ranchers who rent grassland from the government often damage the land and reduce its productive capacity in pursuit of short-term profits.

Response: Change the laws so that people and companies can obtain titles and leases to harvest trees and use grasslands only so long as the health of the biological community is maintained. Eliminate tax subsidies that encourage overexploitation of natural resources, and add tax subsidies for land management that enhances conservation efforts.

Problem: In many countries, governments are inefficient, slow-moving, and bound by excessive regulation, and consequently are ineffective at protecting biological communities.

Response: Local nongovernmental conservation organizations, villages, and neighborhood groups are often the most effective agents for dealing with conservation issues and should be encouraged and supported politically, scientifically, and financially. Bringing together all of the stakeholders for discussions, meetings, and planning sessions is crucial so that local people, in particular, feel that their involvement is important.

Problem: Many businesses, banks, and governments are uninterested in and unresponsive to conservation issues (Figure 5.15).

Response: Lobbying efforts may be effective at changing the policies of institutions that want to avoid bad publicity. Petitions, rallies, letter-writing campaigns, press releases, and economic boycotts all serve a purpose when reasonable requests for change are ignored (Figure 5.16). In many situations, radical environmental groups, such as Greenpeace and EarthFirst!, dominate media attention with dramatic, publicity-grabbing actions, while mainstream conservation organizations, such as The Nature Conservancy and the World Wildlife Fund, follow behind to negotiate a compromise.

"Sir, would you take this latest warning of ecological disaster and pooh-pooh it for me?"

5.15 Business interests and governments are often unwilling to acknowledge and address environmental problems. Citizen activism is sometimes needed to convince businesses and governments that concern for biological diversity often makes economic as well as ecological sense. (© The New Yorker Collection 1992 Dana Fradon from cartoonbank.com. All rights reserved.)

The role of conservation biologists

Conservation biology differs from many other scientific disciplines in that it plays an active role in the preservation of biological diversity in all its forms: species, genetic variability, biological communi-

5.16 Demonstrations, such as this protest against the extensive logging of old-growth forests in the Pacific Northwest, can focus media attention on environmental problems that society must not ignore. This demonstration is peaceful, but other environmental demonstrations have been confrontational and even violent. (Photograph © Michael Graybill and Jan Hodder/Biological Photo Service.)

ties, and ecosystem functions. Members of the diverse disciplines that contribute to conservation biology share the common goal of protecting biological diversity (Norton 1991). The ideas and theories of conservation biology are increasingly being incorporated into political debate, popular culture, and public discussion, and the preservation of biological diversity has been targeted as a priority for new government conservation programs. The goal of conservation biologists is not just to create new knowledge, but to use that knowledge to protect biological diversity (Ehrenfeld 2000).

A broad, thoughtful perspective is necessary to create and continue the most effective conservation programs. In many cases, species are driven to extinction by combinations of factors acting simultaneously or sequentially. Blaming a group of poor, rural people or a certain industry for the destruction of biological diversity is a simplistic and usually ineffective strategy. The challenge is to understand the national and international links that promote the destruction and to find viable alternatives. These alternatives must involve stabilizing the size of the human population, finding a livelihood for rural people in both developing and developed countries that does not damage the environment, providing incentives and penalties that will convince industries to value the environment, and restricting international trade in products that are obtained by damaging the environment. Also important is a willingness on the part of people in developed countries to reduce their consumption of the world's resources and to pay fair prices for products that are produced in a sustainable, nondestructive manner.

If these challenges are to be met, conservation biologists must take on several active roles. First, they must become more effective as *educators*, in the public forum as well as in the classroom. Conservation biologists need to educate as broad a range of people as possible about the problems that stem from the loss of biological diversity (Collett and Karakashain 1996). The efforts of Merlin Tuttle and Bat Conservation International illustrate how public attitudes can be changed. When the government of Austin, Texas, began plans to exterminate hundreds of thousands of Mexican free-tailed bats (*Tadarida brasiliensis*) that lived under a downtown bridge, Tuttle and his colleagues conducted a successful publicity campaign to convince people that bats were both fun to watch and critical in controlling populations of noxious insects over a wide area. The situation has changed so radically that now the government protects the bats as a matter of civic pride, and citizens and tourists gather every night to watch the bats emerge (Figure 5.17).

Second, conservation biologists must become *politically active*. Involvement in the political process allows conservation biologists to influence government policy and propose legislation that would

5.17 Citizens and tourists gather in the evening to watch bats emerge from their roosts on the underside of a bridge in Austin, Texas. (Photograph by Merlin Tuttle, Bat Conservation International.)

prove beneficial to species or ecosystems (Brown 2000). Recent difficulties in getting the U.S. Congress to reauthorize the U.S. Endangered Species Act and to ratify the Convention on Biological Diversity dramatically illustrate the need for scientists to spend more time with politicians explaining the need for conservation.

Third, conservation biologists need to become *organizers* within the biological community. By stimulating interest in conservation biology among their colleagues, conservation biologists can increase the ranks of trained professional advocates fighting the destruction of natural resources.

Fourth, conservation biologists need to become *motivators*, convincing a range of people to support conservation efforts. At a local level, conservation programs have to be created and presented in ways that provide incentives for local people to support them. Local people need to be shown that protecting the environment not only saves species and biological communities, but also improves the long-term health of their families and their own economic well-being (WRI 1998; McMichael et al. 1999). Public discussion, education, and publicity need to be a major part of any such program. Careful attention must be devoted in particular to convincing business leaders and politicians to support conservation efforts. Many of these people will support conservation efforts when they are presented in the right way; sometimes conservation is perceived to have good publicity value, or supporting it is perceived to be better than a confrontation that may otherwise result.

Finally, and most importantly, conservation biologists need to become effective *managers* and *practitioners* of conservation projects. They must be willing to walk on the land and go out on the water to find out what is really happening, to get dirty, to talk and work with local people, to knock on doors, and to take risks. Conservation biologists must learn everything they can about the species and communities that they are trying to protect, and then make that knowledge available to others (Clark 1999; Latta 2000). If conservation biologists are willing to put their ideas into practice, and to work with park managers, land use planners, politicians, and local people, then progress will follow. Getting the right mixture of models, new theories, innovative approaches, and practical examples will be the key to the success of the discipline. Once this balance is found, conservation biologists working with an energized citizenry will be in a position to protect the world's biological diversity during this unprecedented era of change.

Summary

1. Sustainable development has become an important concept in guiding human activities, but it has not often been easy to find the right balance between the protection of biological diversity and the use of natural resources.

2. Local and national governments protect biological diversity through the passage of laws that regulate activities such as fishing and hunting, land use, and industrial pollution, and through the establishment of protected areas.

3. Many traditional societies have strong conservation ethics and management practices that are compatible with the protection of biological diversity, and these people need to be supported in their efforts.

4. Major environmental documents, such as the Convention on Biological Diversity and the Convention on Climate Change, were signed at the 1992 Earth Summit, attended by over 100 heads of state, and at subsequent environmental meetings. Implementing and funding these new treaties could prove vital to international conservation efforts.

5. Conservation groups, governments in developed countries, and development banks are increasing funding to protect biological diversity in developing tropical countries. While the increased levels of funding are welcome, the amount of money is still inadequate to deal with the loss of biological diversity that is taking place. Additional, innovative mechanisms, such as nation-

al environmental funds and debt-for-nature swaps, are being developed to fund conservation activities.

6. International aid agencies and development banks, including the World Bank, have often funded massive projects that cause widespread environmental damage. These funding bodies are now attempting to be more environmentally responsible in their lending policies.

7. Conservation biologists must demonstrate the validity of the theories and approaches of their new discipline, and must actively work with all components of society to protect biological diversity and to restore the degraded elements of the environment.

Suggested Readings

Batisse, M. 1997. Biosphere reserves: A challenge for biodiversity conservation and regional development. *Environment* 39(5): 7–15, 31–35. History and accomplishments of this important program.

Brown, K. 2000. Transforming a discipline: A new breed of scientist-advocate emerges. *Science* 287: 1192–1193. Several articles in this issue describe how conservation biologists and ecologists are working with the government to affect policy.

Chau, K. 1995. The Three Gorges project of China: Resettlement prospects and problems. *Ambio* 24:98–102. The consequences of development on a massive scale. (Take note of this excellent journal on the human environment.)

Collett, J. and S. Karakashain (eds.). 1996. *Greening the College Curriculum: A Guide to Environmental Teaching in the Liberal Arts*. Island Press, Washington, D.C. Environmental and conservation issues can be a theme unifying many university courses and programs of study.

Cox, P. A. 1997. *Nafauna: Saving the Samoan Rain Forest*. W. H. Freeman, New York. Exciting and beautiful account of a scientist's efforts to save a forest and help a village.

Dwyer, J. C. and I. D. Hodge. 1996. *Countryside in Trust: Land Management by Conservation, Recreation and Amenity Organisations*. John Wiley and Sons, Chichester. Description of the enormous growth of land trusts in Britain.

Ehrenfeld, D. 2000. War and peace and conservation biology. *Conservation Biology* 14: 105–112. Unique essay, guaranteed to promote discussion, on the need to evaluate whether conservation biologists are making a difference.

Heywood, V. H. (eds.). 1995. *Global Diversity Assessment*. Cambridge University Press, Cambridge. The most complete single source of information on biological diversity.

Homer-Dixon, T. F. 1999. *Environment, Scarcity and Violence*. Princeton University Press, Princeton. Competition for resources in a degraded environment leads to conflict.

Jacobson, S. K., E. Vaughn and S. W. Miller. 1995. New directions in conservation biology: Graduate programs. *Conservation Biology* 9: 5–17. Descriptions of 51 U.S. graduate programs with lists of faculty.

Kapur, D., J. P. Lewis and R. Webb (eds.). 1997. *The World Bank: Its First Half-Century*. The Brookings Institution, Washington, D.C. Thorough, critical examination of the World Bank's lending policies.

Maser, C. 1997. *Sustainable Community Development: Principles and Practices*. CRC Press, Boca Raton, Florida. Imaginative look at principles of sustainability, drawing on a wide range of examples from around the world.

Meffe, G. K. and C. R. Carroll (eds.). 1997. *Principles of Conservation Biology*, Second Edition. Sinauer Associates, Sunderland, MA. Excellent advanced textbook.

Oates, J. F. 1999. *Myth and Reality in the Rainforest: How Conservation Strategies Are Failing in West Africa*. University of California Press, Berkeley. Many sustainable development projects do not live up to expectations.

Primack, R., D. Bray, H. Galletti and I. Ponciano (eds.). 1998. *Timber, Tourists and Temples: Conservation and Development in the Maya Forest of Belize, Guatemala, and Mexico*. Island Press, Washington, D.C. Intricate social, political, and economic factors affect conservation issues in this important region.

Western, D. 1997. *In the Dust of Kilimanjaro*. Island Press, Washington, D.C. The former Director of the Kenya Wildlife Service provides a personal view of the need for integration of people and wildlife on the African landscape.

Selected Environmental Organizations and Sources of Information

The best single reference on conservation activities is the *Conservation Directory*, published by Lyon Press and updated each year by the National Wildlife Federation, 8925 Leesburg Pike, Vienna, VA 22184. This directory lists thousands of local, national, and international conservation organizations, conservation publications, and leaders in the field of conservation. Other publications of interest include *The Complete Guide to Environmental Careers in the 21st Century* (1998), published by Island Press, 1718 Connecticut Avenue N.W., Suite 300, Washington, D.C. 20009-1148; *Environmental Profiles: A Global Guide to Projects and People* (1993), published by Garland Publishing, 717 Fifth Avenue, New York, NY 10022; and *World's Who's Who and Does What in Environment and Conservation* (1997), by N. Polunin and L. M. Curme, published by St. Martin's Press, New York. Additionally, the *Web of Science* (www.webofscience.com), available through libraries and by subscription, provides a powerful way to search for published research by subject, author, and title.

The following is a list of some major organizations and resources.

American Zoo and Aquarium Association
8403 Colesville Road, Suite 710
Silver Spring, MD 20910-3314 U.S.A.
www.aza.org

Preservation and propagation of captive wildlife.

BirdLife International
Wellbrook Court
Girton Road
Cambridge, CB3 0NA, United Kingdom
www.wing-wbsj.or.jp/birdlife

Formerly known as the International Committee for Bird Protection. Determines conservation status and priorities for birds throughout the world.

Center for Marine Conservation
1725 DeSales St. N.W., Suite 600
Washington, D.C. 20036 U.S.A.
www.cmc-ocean.org

Focuses on marine wildlife and ocean and coastal habitats.

Center for Plant Conservation/
Missouri Botanical Garden
P.O. Box 299
St. Louis, MO 63166-0299 U.S.A.
www.mobot.org

Major centers for worldwide plant conservation activities.

CITES Secretariat, UNEP
International Environment House
15, chemin des Anemones
CH-1219 Châtelaine-Geneva, Switzerland
www.wcmc.org.uk/CITES/eng/index.shtml

Regulates trade in endangered species.

Conservation International
2501 M Street N.W., Suite 200
Washington, D.C. 20037 U.S.A.
www.conservation.org

Active in conservation efforts and in working for sustainable development.

Earthwatch Institute
3 Clocktower Place, Suite 100
P.O. Box 75
Maynard, MA 01754
www.earthwatch.org

Clearinghouse for international projects in which volunteers can work with scientists.

Environmental Data Research Institute
1655 Elmwood Ave., Suite 225
Rochester, NY 14620-3426 U.S.A.

Publishes *Environmental Granting Foundations*, a comprehensive guide to funding sources.

Environmental Defense
257 Park Avenue South
New York, NY 10010 U.S.A.
www.environmentaldefense.org

Involved in scientific, legal, and economic issues.

Friends of the Earth
1025 Vermont Avenue N.W.,
Suite 300
Washington, D.C. 20005-6303 U.S.A.
www.foe.org

International organization working to improve and expand environmental public policy.

Greenpeace U.S.A.
1436 U Street N.W.
Washington, D.C. 20009 U.S.A.
www.greenpeace.org

Activist organization, known for grassroots efforts and dramatic protests against environmental damage.

International Union for the Conservation of
Nature and Natural Resources (IUCN)
Rue Mauverney 28
CH-1196, Gland, Switzerland
www.iucn.org

This is the premier coordinating body for international conservation efforts. Produces directories of specialists who are knowledgeable about captive breeding programs and other aspects of conservation.

National Audubon Society
700 Broadway
New York, N.Y. 10003 U.S.A.
www.audubon.org

Extensive program, including wildlife conservation, public education, research, and political lobbying.

National Council for Science
and the Environment (NCSE)
1725 K Street, N.W., Suite 212
Washington, DC 20006-1401 U.S.A.
www.cnie.org

Formerly the Committee for the National Institute for the Environment. Works to improve the scientific basis for environmental decision making; their web site is an extensive source of environmental information.

National Wildlife Federation
8925 Leesburg Pike
Vienna, VA 22184 U.S.A.
www.nwf.org

Advocates for wildlife conservation. Publishes the *Conservation Directory*, as well as the outstanding children's publications *Ranger Rick* and *Your Big Backyard*.

Natural Resources Defense Council

40 West Twentieth Street
New York, NY 10011 U.S.A.
www.nrdc.org

Uses legal and scientific methods to monitor and influence government actions and legislation.

The Nature Conservancy

International Headquarters
4245 North Fairfax Drive, Suite 100
Arlington, VA 22203-1606 U.S.A.
www.tnc.org

Emphasizes land preservation. Maintains extensive records on rare species distribution in the Americas, particularly North America.

The New York Botanical Garden/ Institute for Economic Botany

200th Street and Kazimiroff Boulevard
Bronx, NY 10458 U.S.A.
www.nybg.org

Conduct research and conservation programs involving plants that are useful to people.

Rainforest Action Network

221 Pine Street, Suite 500
San Francisco, CA 94104 U.S.A.
www.ran.org

Works actively for rain forest conservation.

The Ramsar Convention Bureau

Rue Mauverney 28
CH-1196 Gland, Switzerland
www.iucn.org/themes/ramsar/

Promotes international conservation of wetlands.

Royal Botanic Gardens, Kew

Surrey TW9 3AB, United Kingdom
www.rbgkew.org.uk

The famous "Kew Gardens" are home to a leading botanical research institute and an enormous plant collection.

Sierra Club

85 Second Street, Second Floor
San Francisco, CA 94105-3441 U.S.A.
www.sierraclub.org

Leading advocate for the preservation of wilderness and open space.

Smithsonian Institution/National Zoological Park

100 Jefferson Drive S.W.
Washington, D.C. 20560 U.S.A.
www.si.edu/natzoo

The National Zoo and the nearby U.S. National Museum of Natural History represent a vast resource of literature, biological materials, and skilled professionals.

Society for Conservation Biology

c/o Blackwell Science, Inc.
Commerce Place
350 Main Street
Malden, MA 02148-5018 U.S.A.
www.conbio.rice.edu/scb/

Leading scientific society for the field. Develops and publicizes new ideas and scientific results through the journal *Conservation Biology*.

Student Conservation Association

P.O. Box 550
Charlestown, NH 03603
www.sca-inc.org

Places volunteers with conservation organizations and public agencies.

United Nations Development Programme (UNDP)

One United Nations Plaza
New York, NY 10017 U.S.A.
www.undp.org/indexalt.html

Funds and coordinates international economic development activities, particularly those that use natural resources in a responsible way.

United Nations Environment Programme (UNEP)

1899 F Street N.W.
Washington, D.C. 20006 U.S.A.
www.unep.ch

International program of research and management relating to major environmental problems.

United States Fish and Wildlife Service

Department of the Interior
1849 C Street N.W.
Washington, D.C. 20240 U.S.A.
www.fws.gov

The leading U.S. government agency in the conservation of endangered species, with a vast research and management network. Major activities also take place within other federal government units, such as the National Marine Fisheries Service and the U.S. Forest Service. The Agency for International Development is active in many developing nations. Individual state governments have comparable units, with National Heritage programs being especially relevant. The *Conservation Directory*, mentioned above, shows how these units are organized.

The Wilderness Society
900 Seventeenth Street N.W.
Washington D.C. 20006-2506 U.S.A.
www.tws.org

Organization devoted to preserving wilderness and wildlife.

Wildlife Conservation Society/
New York Zoological Society
2300 Southern Boulevard
Bronx, NY 10460-1099 U.S.A.
www.wcs.org

Leaders in wildlife conservation and research.

World Bank
1818 H Street N.W.
Washington, D.C. 20433 U.S.A.
www.worldbank.org

A multinational bank involved in economic development; increasingly concerned with environmental issues.

World Conservation Monitoring Centre (WCMC)
Information Office
219 Huntingdon Road
Cambridge CB3 0DL, United Kingdom
www.wcmc.org.uk

Monitors global wildlife trade, the status of endangered species, natural resource use, and protected areas.

World Resources Institute (WRI)
10 G Street N.E., Suite 800
Washington, D.C. 20002 U.S.A.
www.wri.org

Research center producing excellent papers on environmental, conservation, and development topics.

World Wildlife Fund (WWF)
1250 Twenty-Fourth Street N.W.
P.O. Box 97180
Washington, D.C. 20077-7180 U.S.A.
www.worldwildlife.org
or www.wwf.org

Major conservation organization, with branches throughout the world. Active both in research and in the management of national parks.

Xerces Society
4828 Southeast Hawthorne Boulevard
Portland, OR 97215-3252 U.S.A.
www.xerces.org

Focuses on the conservation of insects and other invertebrates.

Zoological Society of London
Regent's Park
London NW1 4RY, United Kingdom
www.zsl.org

Center for worldwide activities to preserve nature.

Bibliography

Abramovitz, J. N. 1991. *Investing in Biological Diversity: U.S. Research and Conservation Efforts in Developing Countries*. World Resources Institute, Washington, D.C.

Abramovitz, J. N. 1994. *Trends in Biodiversity Investments: U.S.-Based Funding for Research and Conservation in Developing Countries, 1987–1991*. World Resources Institute, Washington, D.C.

Ackerman, D. 1992. Last refuge of the monk seal. *National Geographic* 181 (January): 128–144.

Agardy, T. S. 1997. *Marine Protected Areas and Ocean Conservation*. R. G. Landes Company, Austin, TX.

Aguirre, A. A. and E. E. Starkey. 1994. Wildlife disease in U.S. National Parks: Historical and coevolutionary perspectives. *Conservation Biology* 8: 654–661.

Akçakaya, H. R., M. A. Burgman and L. R. Ginzburg. 1999. *Applied Population Ecology: Principles and Computer Exercises Using RAMAS® EcoLab*. Sinauer Associates, Sunderland, MA.

Alcock, J. 1993. *Animal Behavior: An Evolutionary Approach*, 5th ed. Sinauer Associates, Sunderland, MA.

Alcorn, J. B. 1984. Development policy, forests and peasant farms: Reflections on Huastec-managed forests' contributions to commercial production and resource conservation. *Economic Botany* 38: 389–406.

Alcorn, J. B. 1991. Ethics, economies and conservation. *In* M. L. Oldfield and J. B. Alcorn (eds.), *Biodiversity: Culture, Conservation and Ecodevelopment*, pp. 317–349. Westview Press, Boulder, CO.

Alcorn, J. B. 1993. Indigenous peoples and conservation. *Conservation Biology* 7: 424–426.

Alford, R. A. and S. J. Richards. 1999. Global amphibian declines: A problem in applied ecology. *Annual Review of Ecology and Systematics* 30: 133–166.

Allan, T. and A. Warren (eds.). 1993. *Deserts, the Encroaching Wilderness: A World Conservation Atlas*. Oxford University Press, London.

Allen, W. H. 1988. Biocultural restoration of a tropical forest: Architects of Costa Rica's emerging Guanacaste National Park plan to make it an integral part of local culture. *BioScience* 38: 156–161.

Allendorf, F. W. and R. F. Leary. 1986. Heterozygosity and fitness in natural populations of animals. *In* M. E. Soulé (ed.), *Conservation Biology: The Science of Scarcity and Diversity*, pp. 57–76. Sinauer Associates, Sunderland, MA.

Altieri, M. A. and M. K. Anderson. 1992. Peasant farming systems, agricultural modernization and the conservation of crop genetic resources in Latin America. *In* P. L. Fiedler and S. K. Jain (eds.), *Conservation Biology: The Theory and Practice of Nature Conservation, Preservation and Management*, pp. 49–64. Chapman and Hall, New York.

Anderson, A. B. (ed.). 1990. *Alternatives to Deforestation*. Columbia University Press, New York.

Anderson, C. W., R. R. Brooks, R. B. Stewart and R. Simcock. 1998. Harvesting a crop of gold in plants. *Nature* 395: 553–554.

Angel, M. V. 1993. Biodiversity of the pelagic ocean. *Conservation Biology* 7: 760–772.

Armbruster, P. and R. Lande. 1993. A population viability analysis for African elephant (*Loxodonta africana*): How big should reserves be? *Conservation Biology* 7: 602–610.

Armstrong, S. and R. Botzler (eds.). 1998. *Environmental Ethics: Divergence and Convergence*. McGraw-Hill, New York.

Avise, J. C. and J. L. Hamrick (eds.). 1996. *Conservation Genetics: Case Histories from Nature*. Chapman and Hall, New York.

Balick, M. J. and P. A. Cox. 1996. *Plants, People and Culture: The Science of Ethnobotany*. Scientific American Library, New York.

Balmford, A. and A. Long. 1994. Avian endemism and forest loss. *Nature* 372: 623–624.

Balmford, A. and K. J. Gaston. 1999. Why biodiversity surveys are good value. *Nature* 398: 204–205.

Baltz, D. M. 1991. Introduced fishes in marine systems and inland seas. *Biological Conservation* 56: 151–177.

Barbier, E. B. 1993. Valuation of environmental resources and impacts in developing countries. *In* R. K. Turner (ed.), *Sustainable Environmental Economics and Management*, pp. 319–337. Belhaven Press, New York.

Barbier, E. B., J. C. Burgess and C. Folke. 1994. *Paradise Lost? The Ecological Economics of Biodiversity*. Earthscan Publications, London.

Barrett, S. C. H. and J. R. Kohn. 1991. Genetic and evolutionary consequences of small population size in plants: Implications for conservation. *In* D. A. Falk and K. E. Holsinger (eds.), *Genetics and Conservation of Rare Plants*, pp. 3–30. Oxford University Press, New York.

Bartley, D., M. Bagley, G. Gall and B. Bentley. 1992. Use of linkage disequilibrium data to estimate effective size of hatchery and natural fish populations. *Conservation Biology* 6: 365–375.

Baskin, Y. 1997. *The Work of Nature: How the Diversity of Life Sustains Us*. Island Press, Washington, D.C.

Baskin, Y. 1998. Winners and losers in a changing world. *BioScience* 48: 788–792.

Batisse, M. 1997. Biosphere reserves: A challenge for biodiversity conservation and regional development. *Environment* 39(5):7–15, 31–35. Bawa, K. S. 1990. Plant-pollinator interactions in tropical rainforests. *Annual Review of Ecology and Systematics* 21: 399–422.

Bawa, K. S. and S. Dayanandan. 1997. Socioeconomic factors and tropical deforestation. *Nature* 386: 562–563.

Bawa, K. S. and S. Menon. 1997. Biodiversity monitoring: The missing ingredients. *Trends in Ecology and Evolution* 12: 42.

Bazzaz, F. A. and E. D. Fajer. 1992. Plant life in a CO_2-rich world. *Scientific American* 266: 68–74.

Beck, B. B. and A. F. Martins. 1995. *Golden Lion Tamarin Reintroduction Annual Report*, 7.

Beck, B. B., L. G. Rapport, M. R. Stanley Price and A. C. Wilson. 1994. Reintroduction of captive-born animals. *In* P. J. Olney, G. M. Mace and A. T. C. Feistner (eds.), *Creative Conservation: Interactive Management of Wild and Captive Animals*, pp. 265–286. Chapman and Hall, London.

Beier, P. and R. F. Noss. 1998. Do habitat corridors provide connectivity? *Conservation Biology* 12: 1241–1252.

Bennett, A. F. 1999. *Linkages in the Landscape: The Role of Corridors and Connectivity in Wildlife Conservation*. IUCN, Gland, Switzerland.

Benz, B. F., L. R. Sánchez-Velásquez and F. J. Santana Michel. 1990. Ecology and ethnobotany of *Zea diploperennis*: Preliminary investigations. *Maydica* 35: 85–98.

Berger, J. 1990. Persistence of different-sized populations: An empirical assessment of rapid extinctions in bighorn sheep. *Conservation Biology* 4: 91–98.

Berger, J. 1999. Intervention and persistence in small populations of bighorn sheep. *Conservation Biology* 13: 432–435.

Bernow, S., et al. 1998. Ecological tax reform. *BioScience* 48: 193–196.

Billington, H. L. 1991. Effect of population size on genetic variation in a dioecious conifer. *Conservation Biology* 5: 115–119.

Birkeland, C. (ed.). 1997. *The Life and Death of Coral Reefs*. Chapman and Hall, New York.

Blaustein, A. R. and D. B. Wake. 1995. The puzzle of declining amphibian populations. *Scientific American* 272 (4): 52–57.

Blomqvist, L. 1995. Three decades of snow leopards, *Panthera uncia*, in captivity. *International Zoo Yearbook* 34: 178–185.

Bodmer, R. E., J. F. Eisenberg and K. H. Redford. 1997. Hunting and the likelihood of extinction of Amazonian mammals. *Conservation Biology* 11: 460–466.

Bowles, I. A. and G. T. Prickett. 1994. *Reframing the Green Window: An Analysis of the GEF Pilot Phase Approach to Biodiversity and Global Warming and Recommendations for the Operational Phase*. Conservation International/National Resources Defense Council, Washington, D.C.

Bowles, M. L. and C. J. Whelan. 1994. *Restoration of Endangered Species: Conceptual Issues, Planning and Implementation*. Cambridge University Press, Cambridge.

Bradshaw, A. D. 1990. The reclamation of derelict land and the ecology of ecosystems. *In* W. R. Jordan III, M. E. Gilpin and J. D. Aber (eds.), *Restoration Ecology: A Synthetic Approach to Ecological Research*, pp. 53–74. Cambridge University Press, Cambridge.

Brandon, K., K. H. Redford and S. E. Sanderson (eds.). 1998. *Parks in Peril: People, Politics and Protected Areas*. Island Press, Washington, D.C.

Brooks, T. and A. Balmford. 1996. Atlantic forest extinctions. *Nature* 380: 115.

Brooks, T. M., S. L. Pimm and J. O. Oyugi. 1999. Time lag between deforestation and bird extinction in tropical forest fragments. *Conservation Biology* 13: 1140–1150.

Brown, G. M. 1993. The economic value of elephants. *In* E. B. Barbier (ed.), *Economics and Ecology: New Frontiers in Sustainable Development*. Chapman and Hall, London.

Brown, K. S. 2000. Transforming a discipline: A new breed of scientist-advocate emerges. *Science* 287: 1192–1193.

Brownlow, C. A. 1996. Molecular taxonomy and the conservation of the red wolf and other endangered carnivores. *Conservation Biology* 10: 390–396.

Brush, S. B. and D. Stabinsky (eds.). 1996. *Valuing Local Knowledge: Indigenous People and Intellectual Property Rights*. Island Press, Washington, D.C.

Bryant, D., D. Nelson and L. Tangley. 1997. *The Last Frontier Forests: Ecosystems and Economies on the Edge*. World Resources Institute, Washington, D.C.

Bryant, D., L. Burke, J. McManus and M. Spalding. 1998. *Reefs at Risk: A Map-Based Indicator of Threats to the World's Coral Reefs*. World Resources Institute, Washington, D.C.

Bryant, E. H., V. L. Backus, M. E. Clark and D. H. Reed. 1999. Experimental tests of captive breeding for endangered species. *Conservation Biology* 13: 1487–1496.

Buchmann, S. L. and G. P. Nabhan. 1996. *The Forgotten Pollinators*. Island Press, Washington, D.C.

Buck, S. J. 1996. *Understanding Environmental Administration and Law*. Island Press, Washington, D.C.

Bulte, E. and G. C. van Kooten. 2000. Economic science, endangered species, and biodiversity loss. *Conservation Biology* 14: 113–119.

Burgman, M. A., S. Ferson and H. R. Akcakaya. 1993. *Risk Assessment in Conservation Biology*. Chapman and Hall, London.

Cairns, J. and J. R. Heckman. 1996. Restoration ecology: The state of an emerging field. *Annual Review of Energy and the Environment* 21: 167–189.

Caldecott, J. 1988. *Hunting and Wildlife Management in Sarawak*. IUCN, Gland, Switzerland.

Caldecott, J. 1996. *Designing Conservation Projects*. Cambridge University Press, Cambridge.

Callicott, J. B. 1990. Whither conservation ethics? *Conservation Biology* 4: 15–20.

Callicott, J. B. 1994. *Earth's Insights: A Multicultural Survey of Ecological Ethics from the Mediterranean Basin to the Australian Outback*. University of California Press, Berkeley, CA.

Carey, A. B. 2000. Effects of new forest management strategies on squirrel populations. *Ecological Applications* 10: 248–257.

Carlton, J. T. and J. B. Geller. 1993. Ecological roulette: the global transport of nonindigenous marine organisms. *Science* 261: 78–82.

Carlton, J. T., J. B. Geller, M. L. Reaka-Kudla and E. A. Norse. 1999. Historical extinctions in the sea. *Annual Review of Ecology and Systematics* 30: 515–538.

Caro, T. (ed.). 1998. *Behavioral Ecology and Conservation Biology*. Oxford University Press, New York.

Caro, T. M. and M. K. Laurenson. 1994. Ecological and genetic factors in conservation: A cautionary tale. *Science* 263: 485–486.

Carroll, C. R. 1992. Ecological management of sensitive natural areas. *In* P. L. Fiedler and S. K. Jain (eds.), *Conservation Biology: The Theory and Practice of Nature Conservation, Preservation and Management*, pp. 347–372. Chapman and Hall, New York.

Carroll, C. R. et al. 1996. Strengthening the use of science in achieving the goals of the Endangered Species Act: An assessment by the Ecological Society of America. *Ecological Applications* 6: 1–12.

Carson, R. 1962. *Silent Spring*. Reprinted in 1982 by Penguin, Harmondsworth, England.

Caswell, H. 1989. *Matrix Population Models: Construction, Analysis and Interpretation*. Sinauer Associates, Inc., Sunderland, MA.

Caughley, G. and A. Gunn. 1996. *Conservation Biology in Theory and Practice*. Blackwell Science, Malden, MA.

Ceballos-Lascuráin, H. (ed.). 1993. *Tourism and Protected Areas*. IUCN, Gland, Switzerland.

Chadwick, D. H. 1993. The American prairie: Roots of the sky. *National Geographic* 184: 90–119.

Chadwick, D. H. 1995. The Endangered Species Act. *National Geographic* 187: 2–41.

Chapin III, F. S., O. E. Sala, E. C. Buke, et al. 1998. Ecosystem consequences of changing biodiversity. *BioScience* 48: 45–52.

Chapin, T. G., D. J. Harrison and D. D. Katnik. 1998. Influence of landscape pattern on habitat use by American marten in an industrial forest. *Conservation Biology* 12: 1327–1337.

Chapman, C. A., S. R. Balcomb, T. R. Gillespie, J. P. Skorupa and T. T. Struhsaker. 2000. Long-term effects of logging on African primate communities: A 28-year comparison from Kibale National Park, Uganda. *Conservation Biology* 14: 207–217.

Chau, K. 1995. The Three Gorges project of China: Resettlement prospects and problems. *Ambio* 24: 98–102.

Cherfas, J. 1991. Disappearing mushrooms: Another mass extinction? *Science* 254: 1458.

Chernela, J. 1987. Endangered ideologies: Tukano fishing taboos. *Cultural Survival Quarterly* 11: 50–52.

Chester, C. C. 1996. Controversy over the Yellowstone's biological resources. *Environment* 38 (6): 10–15, 34–36.

Christensen, N. L., et al. 1996. The report of the Ecological Society of America committee on the scientific basis for ecosystem management. *Ecological Applications* 6: 665–691.

Clark, C. 1992. Empirical evidence for the effect of tropical deforestation on climatic change. *Environmental Conservation* 19: 39–47.

Clark, J. R. 1999. The ecosystem approach from a practical point of view. *Conservation Biology* 13: 679–681.

Clay, J. 1991. Cultural survival and conservation: Lessons from the past twenty years. *In* M. L. Oldfield and J. B. Alcorn (eds.), *Biodiversity: Culture, Conservation and Ecodevelopment*, pp. 248–273. Westview Press, Boulder, CO.

Clemmons, J. R. and R. Buchholz (eds.). 1997. *Behavioral Approaches to Conservation in the Wild*. Cambridge University Press, New York.

Cleveland, D. A., D. Soleri and S. E. Smith. 1994. Do folk crop varieties have a role in sustainable agriculture? *BioScience* 44: 740–751.

Cochrane, M. A., et al. 1999. Positive feedbacks in the fire dynamic of closed canopy tropical forests. *Science* 284: 1832–1835.

Cohn, J. P. 1994. Salamanders: Slip-sliding away or too surreptitious to count? *BioScience* 44: 219–223.

Collett, J. and S. Karakashain (eds.). 1996. *Greening the College Curriculum: A Guide to Environmental Teaching in the Liberal Arts*. Island Press, Washington, D.C.

Colwell, R. K. 1986. Community biology and sexual selection: Lessons from hummingbird flower mites. *In* T. J. Case and J. Diamond (eds.) *Ecological Communities*, pp. 406–424. Harper and Row Publishers, New York.

Condit, R., S. P. Hubbel and R. B. Foster. 1992. Short-term dynamics of a Neotropical forest. *BioScience* 42: 822–828.

Conservation International. 1990. *The Rain Forest Imperative*. Conservation International, Washington, D.C.

Corlett, R. T. and I. M. Turner. 1996. The conservation value of small, isolated fragments of lowland tropical rain forest. *Trends in Ecology and Evolution* 11: 330–333.

Costanza, R., et al. 1998. Principles for sustainable governance of the oceans. *Science* 281: 198–199.

Costanza, R., R. d'Arge, R. de Groot, S. Farber and nine others. 1997. The value of the world's ecosystem services and natural capital. *Nature* 387: 253–260.

Couzin, J. 1999. Landscape changes make regional climate run hot and cold. *Science* 283: 317-318.

Cowlishaw, G. 1999. Predicting the pattern of decline of African primate diversity: An extinction debt from historical deforestation. *Conservation Biology* 13: 1183–1193.

Cox, G. W. 1993. *Conservation Ecology*. W. C. Brown, Dubuque, IA.

Cox, P. A. 1997. *Nafanua: Saving the Samoan Rain Forest*. W. H. Freeman, New York.

Cox, P. A. and T. Elmqvist. 1997. Ecocolonialism and indigenous-controlled rainforest preserves in Samoa. *Ambio* 26: 84–89.

Cox, P. A., T. Elmqvist, D. Pierson and W. E. Rainey. 1991. Flying foxes as strong interactors in South Pacific island ecosystems: A conservation hypothesis. *Conservation Biology* 5: 448–454.

Crumpacker, D. W., S. W. Hodge, D. Friedley and W. P. Gregg, Jr. 1988. A preliminary assessment of the status of major terrestrial and wetland ecosystems on federal and Indian lands in the United States. *Conservation Biology* 2: 103–115.

Curio, E. 1996. Conservation needs ethology. *Trends in Ecology and Evolution* 11: 260–263.

Currie, D. J. 1991. Energy and large-scale patterns of animal- and plant-species richness. *American Naturalist* 137: 27–49.

Daily, G. C. (ed.). 1997. *Nature's Services: Societal Dependence on Ecosystem Services*. Island Press, Washington, D.C.

Daily, G. C. 1995. Restoring value to the world's degraded lands. *Science* 269: 350–354.

Dalton, R. 1999. Monitoring system planned for U.S. biodiversity drive. *Nature* 398: 738.

Daly, H. E. and J. B. Cobb Jr. 1989. *For the Common Good: Redirecting the Economy Toward Community, the Environment, and a Sustainable Future*. Beacon Press, Boston MA.

Dasgupta, P. S. 1995. Population, poverty and the local environment. *Scientific American* 272: 40–45.

Dasmann, R. F. 1991. The importance of cultural and biological diversity. *In* M. L. Oldfield and J. B. Alcorn (eds.), *Biodiversity: Culture, Conservation and Ecodevelopment*, pp. 7–15. Westview Press, Boulder, CO.

Dasmann, R. F., J. P. Milton and P. H. Freeman. 1973. *Ecological Principles for Economic Development*. John Wiley & Sons, London.

Daszak, P., A. A. Cunningham and A. D. Hyatt. 2000. Emerging infectious diseases of wildlife—threats to biodiversity and human health. *Science* 287: 443–449.

Davis, M. B. and C. Zabinski. 1992. Changes in geographical range resulting from greenhouse warming: Effects on biodiversity in forests. *In* R. Peters and T. E. Lovejoy. (eds.), *Global Warming and Biological Diversity*, pp. 297–308. Yale University Press, New Haven, CT.

Davis, S. D. et al. 1986. *Plants In Danger: What Do We Know?* IUCN, Gland, Switzerland.

Davis, W. 1995. *One River: Explorations and Discoveries in the Amazon Rainforest.* Simon & Schuster, New York.

Del Tredici, P. 1991. Ginkgos and people: A thousand years of interaction. *Arnoldia* 51: 2–15.

DeMauro, M. M. 1993. Relationship of breeding system to rarity in the lakeside daisy (*Hymenoxys acaulis* var. *glabra*). *Conservation Biology* 7: 542–550.

Denslow, J. S. and C. Padoch, (eds.). 1988. *People of the Tropical Rain Forest.* University of California Press, Berkeley, CA.

Devall, B. and G. Sessions. 1985. *Deep Ecology: Living as if Nature Mattered.* Gibbs Smith Publishers, Salt Lake City, UT.

Diamond, A. W. 1985. The selection of critical areas and current conservation efforts in tropical forest birds. *In* A. W. Diamond and T. E. Lovejoy (eds.), *Conservation of Tropical Forest Birds*, pp. 33–48. Technical Publication No. 4, International Council for Bird Preservation, Cambridge, UK.

Diamond, J. M. 1975. The island dilemma: Lessons of modern biogeographic studies for the design of natural reserves. *Biological Conservation* 7: 129–146.

Diamond, J. M. 1999. Dirty eating for healthy living. *Nature* 400: 120–121.

Dietz, J. M., L. A. Dietz and E. Y. Nagagata. 1994. The effective use of flagship species for conservation of biodiversity: The example of lion tamarins in Brazil. *In* P. J. S. Olney, G. M. Mace and A. T. C. Feistner (eds.), *Creative Conservation: Interactive Manangement of Wild and Captive Animals*, Chapman and Hall, London.

Dinerstein, E. and G. F. McCracken. 1990. Endangered greater one-horned rhinoceros carry high levels of genetic variation. *Conservation Biology* 4: 417–422.

Dobson, A. 1995. Biodiversity and human health. *Trends in Ecology and Evolution* 10: 390–392.

Dobson, A. 1998. *Conservation and Biodiversity.* Scientific American Library, No. 59. W. H. Freeman and Co., New York.

Dobson, A. P., J. P. Rodriguez, W. M. Roberts and D. S. Wilcove. 1997a. Geographic distribution of endangered species in the United States. *Science* 275: 550–554.

Dobson, A. P., A. D. Bradshaw and A. J. M. Baker. 1997b. Hopes for the future: Restoration ecology and conservation biology. *Science* 277: 515–522.

Dodd, C. K. and R. A. Seigel. 1991. Relocation, repatriation and translocation of amphibians and reptiles: Are they conservation strategies that work? *Herpetologica* 47: 336–350.

Donahue, D. L. 1999. *Western Range Revisited: Removing Livestock from Public Lands to Conserve Native Biodiversity.* University of Oklahoma Press, Norman, OK.

Dowling, T. E. and M. R. Childs. 1992. Impact of hybridization on a threatened trout of the south-western United States. *Conservation Biology* 6: 355–364.

Drake, J. A., et al. (eds.). 1989. *Biological Invasions: A Global Perspective.* SCOPE Report No. 37. John Wiley, New York.

Drayton, B. and R. Primack. 1996. Plant species lost in an isolated conservation area in metropolitan Boston from 1894 to 1993. *Conservation Biology* 10: 30–40.

Dregné, H. E. 1983. *Desertification of Arid Lands.* Academic Press, New York.

Duffus, D. A. and P. Dearden. 1990. Non-consumptive wildlife-oriented recreation: A conceptual framework. *Biological Conservation* 53: 213–231.

Duffy, E. and A. S. Watts (eds.). 1971. *The Scientific Management of Animal and Plant Communities for Conservation.* Blackwell Scientific Publications, Oxford.

Dugan, P. (ed.). 1993. *Wetlands in Danger: A World Conservation Atlas.* Oxford University Press, New York.

Dwyer, J. C. and I. D. Hodge. 1996. *Countryside in Trust: Land Management by Conservation, Recreation and Amenity Organizations.* John Wiley and Sons, Chichester, UK.

Eales, S. 1992. *Earthtoons: The First Book of Ecohumor.* Warner Books, New York.

Easter-Pilcher, A. 1996. Implementing the Endangered Species Act: Assessing the listing of species as endangered or threatened. *BioScience* 46: 355–363.

Ehrenfeld, D. W. 1989. Hard times for diversity. *In* D. Western and M. Pearl (eds.), *Conservation for the Twenty-first Century*, pp. 247–250. Oxford University Press, New York.

Ehrenfeld, D. W. 2000. War and peace and conservation biology. *Conservation Biology* 14: 105–112.

Eisner, T. 1991. Chemical prospecting: A proposal for action. *In* F. H. Bormann and S. R. Kellert (eds.), *Ecology, Economics, Ethics: The Broken Circle*, pp. 196–202. Yale University Press, New Haven, CT.

Eisner, T. and E. A. Beiring. 1994. Biotic exploration fund: Protecting biodiversity through chemical prospecting. *BioScience* 44: 95–98.

Elfring, C. 1989. Preserving land through local land trusts. *BioScience* 39: 71–74.

Ellstrand, N. C. 1992. Gene flow by pollen: Implications for plant conservation genetics. *Oikos* 63: 77–86.

Endangered Species Coalition. 1992. *The Endangered Species Act: A Commitment Worth Keeping.* The Wilderness Society, Washington, D.C.

Enderson, J. H., et al. 1995. Population changes in North American peregrines. *Transactions of the 60th North American Wildlife and Natural Resource Conference*, 142–161.

Enserink, M. 1999. Biological invaders sweep in. *Science* 285: 1834–1836.

Erdelen, W. 1988. Forest ecosystems and nature conservation in Sri Lanka. *Biological Conservation* 43: 115–135.

Falk, D. A. 1991. Joining biological and economic models for conserving plant genetic diversity. *In* D. A. Falk and K. E. Holsinger (eds.), *Genetics and Conservation of Rare Plants*, pp. 209–224. Oxford University Press, New York.

Falk, D. A. and K. E. Holsinger (eds.). 1991. *Genetics and Conservation of Rare Plants.* Oxford University Press, New York.

Falk, D. A. and P. Olwell. 1992. Scientific and policy considerations in restoration and reintroduction of endangered species. *Rhodora* 94: 287–315.

Falk, D. A., C. I. Millar and M. Olwell (eds.). 1996. *Restoring Diversity: Strategies for Reintroduction of Endangered Plants.* Island Press, Washington, D.C.

Farnsworth, E. J. and J. Rosovsky. 1993. The ethics of ecological field experimentation. *Conservation Biology* 7: 463–472.

Farnsworth, N. R. 1988. Screening plants for new medicines. *In* E. O. Wilson and F. M. Peter (eds.), *Biodiversity*, pp. 83–97. National Academy Press, Washington, D.C.

Fearnside, P. M. 1990. Predominant land uses in Brazilian Amazonia. *In* A. Anderson (ed.), *Alternatives to Deforestation: Steps Toward Sustainable Use of the Amazon Rain Forest*, pp. 233–251. Columbia University Press, Irvington, NY.

Fillon, F. L., A. Jacquemot and R. Reid. 1985. *The Importance of Wildlife to Canadians.* Canadian Wildlife Service, Ottawa.

Finlay, B. J. and K. J. Clarke. 1999. Ubiquitous dispersal of microbial species. *Nature* 400: 828.

Fisk, M. R., S. J. Giovannoni and I. H. Thorseth. 1998. Alteration of oceanic volcanic glass: Textural evidence of microbial activity. *Science* 281: 978–980.

Flather, C. H., M. S. Knowles and I. A. Kendall. 1998. Threatened and endangered species geography. *BioScience* 48: 365–376.

Fleischner, T. L. 1994. Ecological costs of livestock grazing in western North America. *Conservation Biology* 8: 629–644.

Foin, T. C., et al. 1998. Improving recovery planning for threatened and endangered species. *BioScience* 48: 177–184.

Foose, T. J. 1983. The relevance of captive populations to the conservation of biotic diversity. *In* C. M. Schonewald-Cox, S. M. Chambers, B. MacBryde and L. Thomas (eds.). *Genetics and Conservation*, pp. 374–401. Benjamin/Cummings, Menlo Park, CA.

Forman, R. T. 1995. *Land Mosaics: The Ecology of Landscapes and Regions.* Cambridge University Press, New York.

Frankham, R. 1996. Relationships of genetic variation to population size in wildlife. *Conservation Biology* 10: 1500–1508.

Franklin, I. R. 1980. Evolutionary change in small populations. *In* M. E. Soulé and B. A. Wilcox (eds.). *Conservation Biology: An Evolutionary-Ecological Perspective*, pp. 135–149. Sinauer, Sunderland, MA.

Frederick, R. J. and M. Egan. 1994. Environmentally compatible applications of biotechnology. *BioScience* 44: 529–535.

Fredrickson, J. K. and T. C. Onstatt. 1996. Microbes deep inside Earth. *Scientific American* 275: 68–73.

Freese, C. H. (ed.). 1997. *Harvesting Wild Species: Implications for Biodiversity Conservation.* Johns Hopkins University Press, Baltimore, MD.

French, H. F. 1994. Making environmental treaties work. *Scientific American* 271: 94–97.

Fuccilo, D., L. Sears and P. Stapleton. 1998. *Biodiversity in Trust: Conservation and Use of Plant Genetic Resources in CGIAR Centres.* Cambride University Press, New York.

Fujita, M. S. and M. D. Tuttle. 1991. Flying foxes (Chiroptera: Pteropodidae): Threatened animals of key ecological and economic importance. *Conservation Biology* 5: 455–463.

Fuller, R. J., R. D. Gregory, D. W. Gibbons, J. H. Marchant, et al. 1995. Population declines and range contractions among lowland farmland birds in Britain. *Conservation Biology* 9: 1425–1441.

Funch, P. and R. Kristensen. 1995. Cycliophora is a new phylum with affinities to Entoprocta and Ectoprocta (*Symbion pandora*). *Nature* 378: 711–714.

Futuyma, D. J. 1986. *Evolutionary Biology*, Second Edition. Sinauer Associates, Sunderland, MA.

Gadgil, M. and R. Guha. 1992. *This Fissured Land: An Ecological History of India.* Oxford University Press, Oxford.

Galatowitsch, S. M. and A. G. Van der Valk. 1996. *Restoring Prairie Wetlands: An Ecological Approach.* Iowa State University Press, Ames, IA.

Gates, D. M. 1993. *Climate Change and Its Biological Consequences.* Sinauer Associates, Sunderland, MA.

Gentry, A. H. 1986. Endemism in tropical versus temperate plant communities. *In* M. E. Soulé (ed.), *Conservation Biology: The Science of Scarcity and Diversity*, pp. 153–181. Sinauer Associates, Sunderland, MA.

Gerrodette, T. and W. G. Gilmartin. 1990. Demographic consequences of changing pupping and hauling sites of the Hawaiian monk seal. *Conservation Biology* 4: 423–430.

Gersh, J. and R. Pickert. 1991. Land-use modeling: Accommodating growth while conserving biological resources in Dutchess County, New York. *In* D. J. Decker, M. E. Krasnyk, G. R. Goff, C. R. Smith and D. W. Gross (eds.), *Challenges in the Conservation of Biological Resources: A Practitioner's Guide*, pp. 233–242. Westview Press, Boulder, CO.

Getz, W. M., et al. 1999. Sustaining natural and human capital: Villagers and scientists. *Science* 283: 1855–1856.

Giese, M. 1996. Effects of human activity on adelie penguin *Pygoscelis adeliae* breeding success. *Biological Conservation* 75: 157–164.

Gigon, A., R. Langenauer, C. Meier and B. Nievergelt. 1998. Blaue Listen der erfolgreich erhaltenen oder geförderten Tier- und Pflanzenarten der Roten Listen. Methodik und Anwendung in der nördlichen Schweiz. *Veröff. Geobot. Inst.* 129: 1–137. ETH, Stiftung Rübel, Zürich.

Gilbert, O. L. and P. Anderson. 1998. *Habitat Creation and Repair.* Oxford University Press, Oxford.

Gilpin, M. E. and M. E. Soulé. 1986. Minimum viable populations: Processes of species extinction. *In* M. E. Soulé (ed.), *Conservation Biology: The Science of Scarcity and Diversity*, pp. 19–34. Sinauer Associates, Sunderland, MA.

Giovannoni, S. J., T. B. Britschgi, C. L. Moyer and K. G. Field. 1990. Genetic diversity in Sargasso Sea bacterioplankton. *Nature* 345: 60–63.

Gipps, J. H. W. (ed.). 1991. *Beyond Captive Breeding: Reintroducing Endangered Species Through Captive Breeding.* Zoological Society of London Symposia, No. 62. Clarendon Press, Oxford.

Gittleman, J. L. 1994. Are the pandas successful specialists or evolutionary failures? *BioScience* 44: 456–464.

Given, D. 1994. *Principles and Practices of Plant Conservation*. Timber Press, Portland, OR.

Godoy, R. A., R. Lubowski and A. Markandya. 1993. A method for the economic valuation of non-timber tropical forest products. *Economic Botany* 47: 220–233.

Goklany, I. M. 1998. Saving habitat and conserving biodiversity on a crowded planet. *BioScience* 48: 941–953.

Goldammer, J. G. 1999. Forests on fire. *Science* 284: 1782–1783.

Goldsmith, B. (ed.). 1991. *Monitoring for Conservation and Ecology*. Chapman and Hall, New York.

Gomez-Pompa, A. and A. Kaus. 1992. Taming the wilderness myth. *BioScience* 42: 271–279.

Goodland, R. J. A. 1990. The World Bank's new environmental policy for dams and reservoirs. *Water Resources Development* 6: 226–239.

Grant, P. R. and B. R. Grant. 1992. Darwin's finches: Genetically effective population sizes. *Ecology* 73: 766–784.

Grassle, J. F., P. Lasserre, A. D. McIntyre and G. C. Ray. 1991. Marine biodiversity and ecosystem function. *Biology International*. Special Issue 23: i-iv, 1–19.

Green, G. N. and R. W. Sussman. 1990. Deforestation history of the eastern rain forests of Madagascar from satellite images. *Science* 248: 212–215.

Gregg, W. P., Jr. 1991. MAB Biosphere Reserves and conservation of traditional land use systems. *In* M. L. Oldfield and J. B. Alcorn (eds.), *Biodiversity: Culture, Conservation and Ecodevelopment*, pp. 274–294. Westview Press, Boulder, CO.

Griffith, B., J. M. Scott, J. W. Carpenter and C. Reed. 1989. Translocation as a species conservation tool: Status and strategy. *Science* 245: 477–480.

Grifo, F. and J. Rosenthal (eds.). 1997. *Biodiversity and Human Health*. Island Press, Washington, D.C.

Grumbine, E. R. 1994a. What is ecosystem management? *Conservation Biology* 8: 27–38.

Grumbine, E. R. 1994b. *Environmental Policy and Biodiversity*. Island Press, Washington, D.C.

Guerrant, E. O. 1992. Genetic and demographic considerations in the sampling and reintroduction of rare plants. *In* P. L. Fiedler and S. K. Jain (eds.), *Conservation Biology: The Theory and Practice of Nature Conservation, Preservation and Management*, pp. 321–344. Chapman and Hall, New York.

Guerrant, E. O. and B. M. Pavlik. 1998. Reintroduction of rare plants: Genetics, demography and the role of *ex-situ* conservation methods. *In* P. L. Fiedler and P. M. Kareiva (eds.), *Conservation Biology: For the Coming Decade*. Chapman and Hall, New York.

Gupta, T. A. and A. Guleria. 1982. *Non-wood Forest Products from India*. IBH Publishing Co., New Delhi.

Haas, P. M., M. A. Levy and E. A. Parson. 1992. Appraising the Earth Summit: How should we judge UNCED's success? *Environment* 34 (8): 7–35.

Hackel, J. D. 1999. Community conservation and the future of Africa's wildlife. *Conservation Biology* 13: 726–734.

Hails, A. J. (ed.). 1996. *Wetlands, Biodiversity and the Ramsar Convention: The Role of the Convention on Wetlands in the Conservation and Wise Use of Biodiversity*. Ramsar Convention Bureau, Gland, Switzerland.

Hall, S. J. G. and J. Ruane. 1993. Livestock breeds and their conservation: A global overview. *Conservation Biology* 7: 815–826.

Halliday, T. 1998. A declining amphibian conundrum. *Nature* 394: 418–419.

Halvorson, W. L. and G. E. Davis (eds.). 1996. *Science and Ecosystem Management in the National Parks*. University of Arizona, Tucson.

Hammond, P. M. 1992. Species Inventory. *In* WCMC, *Global Diversity: Status of the Earth's Living Resources*. pp. 17–39. Chapman and Hall, London.

Hamrick, J. L. and M. J. W. Godt. 1989. Allozyme diversity in plant species. *In* A. H. D. Brown, M. T. Clegg, A. L. Kahler and B. S. Weir (eds.), *Plant Population Genetics, Breeding, and Genetic Resources*, pp. 43–63. Sinauer Associates, Sunderland, MA.

Hansen, S. 1989. Debt for nature swaps—overview and discussion of key issues. *Ecological Economics* 1: 77–93.

Hanski, I. and D. Simberloff. 1997. The metapopulation approach, its history, conceptual domain and application to conservation. *In* I. Hanski and D. Simberloff (eds.), *Metapopulation Biology*, pp. 5–26. Academic Press, Inc., San Diego, CA.

Hanski, I., A. Moilanen and M. Gyllenberg. 1996. Minimum viable metapopulation size. *American Naturalist* 147: 527–541.

Hansson, L., L. Fahrig and G. Merriam (eds.). 1995. *Mosaic Landscapes and Ecological Processes*. Chapman and Hall, London.

Hardin, G. 1985. *Filters Against Folly: How to Survive Despite Economists, Ecologists and the Merely Eloquent*. Viking Press, New York.

Hardin, G. 1993. *Living Within Limits: Ecology, Economics and Population Taboos*. Oxford University Press, New York.

Hargrove, E. C. 1989. *Foundations of Environmental Ethics*. Prentice-Hall, Englewood Cliffs, NJ.

Harrison, J. L. 1968. The effect of forest clearance on small mammals. *In Conservation in Tropical Southeast Asia*. IUCN, Morges, Switzerland.

Hawksworth, D. L. 1990. The long-term effects of air pollutants on lichen communities in Europe and North America. *In* G. M. Woodwell (ed.), *The Earth in Transition: Patterns and Processes of Biotic Impoverishment*, pp. 45–64. Cambridge University Press, Cambridge.

Hellmann, J. J. and G. W. Fowler. 1999. Bias, precision, and accuracy of four measures of species richness. *Ecological Applications* 9: 824–834.

Hemley, G. (ed.). 1994. *International Wildlife Trade: A CITES Sourcebook*. Island Press, Washington, D.C.

Hendrickson, D. A. and J. E. Brooks. 1991. Transplanting short-lived fishes in North American deserts: Review, assessment and recommendations. *In* W. L. Minckley and J. E. Deacon (eds.), *Battle Against Extinction: Native Fish Management in the American West*. University of Arizona Press, Tuscon.

Herren, H. R. and P. Neuenschwander. 1991. Biological control of cassava pests in Africa. *Annual Review of Entomology* 36: 257–283.

Heschel, M. S. and K. N. Paige. 1995. Inbreeding depression, environmental stress and population size variation in Scarlet Gilia (*Ipomopsis aggregata*). *Conservation Biology* 9: 126–133.

Heywood, V. H. (ed.). 1995. *Global Diversity Assessment*. Cambridge University Press, Cambridge.

Heywood, V. H., G. M. Mace, R. M. May and S. N. Stuart. 1994. Uncertainties in extinction rates. *Nature* 368: 105.

Hoffman, A. J., M. H. Bazerman and S. L. Yaffee 1997. Balancing business interests and endangered species protection. *Sloan Management Review* 39 (1): 59–73.

Holloway, M. 1994. Nurturing nature. *Scientific American* 270: 98–108.

Homer-Dixon, T. F. 1999. *Environment, Scarcity, and Violence*. Princeton University Press, Princeton, NJ.

Horwich, R. H. and J. Lyon. 1998. Community-based development as a conservation tool: The Community Baboon Sanctuary and the Gales Point Manatee Reserve. *In* R. B. Primack, D. Bray, H. A. Galletti and I. Ponciano (eds.), *Timber, Tourists, and Temples: Conservation and Development in the Maya Forest of Belize, Guatemala, and Mexico*, pp. 343–364. Island Press, Washington, D.C.

Howard, P. C., et al. 1998. Complementarity and the use of indicator groups for reserve selection in Uganda. *Nature* 394: 472–474.

Hunter, M. L., Jr. 1999. *Maintaining Biodiversity in Forested Ecosystems*. Cambridge University Press, New York.

Huston, M. A. 1994. *Biological Diversity: The Coexistence of Species on Changing Landscapes*. Cambridge University Press, Cambridge.

Hutchings, M. J. 1987. The population biology of the early spider orchid, *Ophrys sphegodes* Mill. I. A demographic study from 1975–1984. *Journal of Ecology* 75: 711–727.

Iltis, H. H. 1988. Serendipity in the exploration of biodiversity: What good are weedy tomatoes? *In* E. O. Wilson and F. M. Peter (eds.), *Biodiversity*, pp. 98–105. National Academy Press, Washington, D.C.

Intergovernmental Panel on Climate Change (IPCC). 1996. *Climate Change 1995: The Science of Climate Change*. World Meteorological Organization and United Nations Environmental Program.

IUCN. 1990. *The IUCN Red Data Book.* IUCN, Gland, Switzerland.

IUCN. 1994a. *Guidelines for Protected Area Management Categories.* IUCN, Gland, Switzerland.

IUCN. 1994b. *IUCN Red List Categories.* IUCN, Gland, Switzerland.

IUCN. 1996. *1996 IUCN Red List of Threatened Animals.* IUCN, Gland, Switzerland.

IUCN/TNC/WWF. 1994. *Report of the First Global Forum on Environmental Funds.* IUCN, Gland, Switzerland.

IUCN/UNEP. 1988. *Coral Reefs of the World.* 3 Volumes. IUCN, Gland, Switzerland.

IUCN/UNEP/WWF. 1991. *Caring for the Earth: A Strategy for Sustainable Living.* Gland, Switzerland.

Jacobson, G. L., Jr., H. Almquist-Jacobson and J. C. Winne. 1991. Conservation of rare plant habitat: Insights from the recent history of vegetation and fire at Crystal Fen, northern Maine, USA. *Biological Conservation* 57: 287–314.

Jacobson, S. K., E. Vaughn and S. W. Miller. 1995. New directions in conservation biology: Graduate programs. *Conservation Biology* 9: 5–17.

Jaffe, M. 1994. *And No Birds Sing.* Simon and Schuster, New York, NY.

James, F. C., C. E. McCulloch and D. A. Wiedenfeld. 1996. New approaches to the analysis of population trends in land birds. *Ecology* 77: 13–27.

Janzen, D. H. 1986. The eternal external threat. *In* M. Soulé (ed.), *Conservation Biology: The Science of Scarcity and Diversity*, pp. 286–303. Sinauer Associates, Sunderland, MA.

Janzen, D. H. 1988a. Tropical dry forests: The most endangered major tropical ecosystem. *In* E. O. Wilson and F. M. Peter (eds.), *Biodiversity.* National Academy Press, Washington, D.C.

Janzen, D. H. 1988b. Tropical ecological and biocultural restoration. *Science* 239: 243–244.

Janzen, D. H. 1999. Gardenification of tropical conserved wildlands: Multitasking, multicropping, and multiusers. *Proceedings of the National Academy of Sciences of the U.S.A.* 96 (11): 5987–5994.

Janzen, D. H. 2000. How to grow a wildland: The gardenification of nature. *In* P. H. Raven and T. Williams (eds.), *Nature and Human Society*, pp. 521–529. National Academy Press, Washington, D.C.

Jarrell, K. R., D. P. Bayley, J. D. Correia and N. A. Thomas. 1999. Recent excitement about the Archaea. *BioScience* 49: 530–541.

Jenkins, R. E. 1996. Natural Heritage Data Center Network: Managing information for managing biodiversity. *In* R. C. Szaro and D. W. Johnston (eds.), *Biodiversity in Managed Landscapes: Theory and Practice*, pp. 176–192. Oxford University Press, New York.

Jiménez, J. A., K. A. Hughes, G. Alaks, L. Graham and R. C. Lacy. 1994. An experimental study of inbreeding depression in a natural habitat. *Science* 266: 271–273.

Johns, A. D. 1987. The use of primary and selectively logged rainforest by Malaysian Hornbills (Bucerotidae) and implications for their conservation. *Biological Conservation* 40: 179–190.

Johnson, N. 1995. *Biodiversity in the Balance: Approaches to Setting Geographic Conservation Priorities.* Biodiversity Support Program, World Wildlife Fund, Washington, D.C.

Jones, H. L. and J. M. Diamond. 1976. Short-time-base studies of turnover in breeding birds of the California Channel Islands. *Condor* 76: 526–549.

Jordan, W. R., III, M. E. Gilpin and J. D. Aber (eds.). 1990. *Restoration Ecology: A Synthetic Approach to Ecological Research.* Cambridge University Press, Cambridge.

Kapur, D., J. P. Lewis and R. Webb (eds.). 1997. *The World Bank: Its First Half-Century.* The Brookings Institution, Washington, D.C.

Karr, J. R. and E. W. Chu. 1998. *Restoring Life in Running Waters.* Island Press, Washington, D.C.

Kaufman, L. 1992. Catastrophic change in a species-rich freshwater ecosystem: Lessons from Lake Victoria. *BioScience* 42: 846–858.

Kaufman, L. and A. S. Cohen. 1993. The great lakes of Africa. *Conservation Biology* 7: 632–633.

Keller, L. F., P. Arcese, J. N. M. Smith, W. M. Hochachka and S. C. Stearns. 1994. Selection against inbred song sparrows during a natural population bottleneck. *Nature* 372: 356–357.

Kellert, S. R. and E. O. Wilson (eds.). 1993. *The Biophilia Hypothesis*. Island Press, Washington, D.C.

Kerr, R. A. 1998. Acid rain control: Success on the cheap. *Science* 282: 1024–1027.

Kiew, R. 1991. *The State of Nature Conservation in Malaysia*. Malayan Nature Society, Kuala Lumpur.

Kimura, M. and J. F. Crow. 1963. The measurement of effective population numbers. *Evolution* 17: 279–288.

Kinnaird, M. F. and T. G. O'Brien. 1991. Viable populations for an endangered forest primate, the Tana River crested mangabey (*Cercocebus galeritus galeritus*). *Conservation Biology* 5: 203–213.

Kleiman, D. G. 1989. Reintroduction of captive mammals for conservation. *BioScience* 39: 152–161.

Kline, V. M. and E. A. Howell. 1990. Prairies. *In* W. R. Jordan III, M. E. Gilpin and J. D. Aber (eds.), *Restoration Ecology: A Synthetic Approach to Ecological Research*, pp. 75–84. Cambridge University Press, Cambridge.

Kloor, K. 2000. Returning America's forests to their "natural" roots. *Science* 287: 573–574.

Kohm, K. and J. F. Franklin (eds.). 1997. *Creating a Forestry for the 21st Century: The Science of Ecosystem Management*. Island Press, Washington, D.C.

Koopowitz, H., A. D. Thornhill and M. Andersen. 1994. A general stochastic model for the prediction of biodiversity losses based on habitat conversion. *Conservation Biology* 8: 425–438.

Kothari, A., N. Singh and S. Suri (eds.). 1996. *People and Protected Areas: Toward Participatory Conservation in India*. Sage Publications, New Delhi.

Kozol, A. et al. 1994. Genetic variation in the endangered burying beetle *Nicrophorus americanus* (Coleoptera: Silphidae). *Annals of the Entomological Society of America* 87: 928–935.

Kramer, R., C. van Shaik and J. Johnson (eds.). 1997. *Last Stand: Protected Areas and Defense of Tropical Biodiversity*. Oxford University Press, New York.

Krebs, J. R., J. D. Wilson, R. B. Bradbury and B. M. Siriwardena. 1999. The second silent spring? *Nature* 400: 611–612.

Kremen, C., A. M. Merenlender and D. D. Murphy. 1994. Ecological monitoring: A vital need for integrated conservation and development programs in the tropics. *Conservation Biology* 8: 388–397.

Kremen, C., et al. 1999. Designing the Masoala National Park in Madagascar based on biological and socioeconomic data. *Conservation Biology* 13: 1055–1068.

Kricher, J. 1998. *A Neotropical Companion*. Princeton University Press, Princeton, NJ.

Kristensen, R. M. 1983. Loricifera, a new phylum with Aschelminthes characters from the meiobenthos. *Zeitschrift fur Zoologische Systematik* 21: 163–180.

Lacy, R. C. and D. B. Lindenmayer. 1995. A simulation study of the impacts of population subdivision on the mountain brushtail possum *Trichosurus caninus* Ogilby (Phalangeridae: Marsupialia), in south-eastern Australia: Loss of genetic variation within and between subpopulations. *Biological Conservation* 73: 131–142.

Lande, R. 1988. Genetics and demography in biological conservation. *Science* 241: 1455–1460.

Lande, R. 1995. Mutation and conservation. *Conservation Biology* 9: 782–792.

Lasiak, T. 1991. The susceptibility and/or resilience of rocky littoral molluscs to stock depletion by the indigenous coastal people of Transkei, southern Africa. *Biological Conservation* 56: 245–264.

Latta, S. C. 2000. Making the leap from researcher to planner: Lessons from avian conservation planning in the Dominican Republic. *Conservation Biology* 14: 132–139.

Laurance, W. F. and R. O. Bierregaard, Jr. (eds.). 1997. *Tropical Forest Remnants: Ecology, Management and Conservation of Fragmented Communities*. The University of Chicago Press, Chicago.

Lawton, J. H. and R. M. May (eds.). 1995. *Extinction Rates*. Oxford University Press, Oxford.

Leader-Williams, N. 1990. Black rhinos and African elephants: Lessons for conservation funding. *Oryx* 24: 23–29.

Ledig, F. T. 1988. The conservation of diversity in forest trees. *BioScience* 38: 471–479.

Lehmkuhl, J. F., R. K. Upreti and U. R. Sharma. 1988. National parks and local development: Grasses and people in Royal Chitwan National Park, Nepal. *Environmental Conservation* 15: 143–148.

Lesica, P. and F. W. Allendorf. 1992. Are small populations of plants worth preserving? *Conservation Biology* 6: 135–139.

Levitus, S., J. I. Antonov, T. P. Boyer and C. Stephens. 2000. Warming of the world ocean. *Science* 287: 2225–2229.

Lewis, D. M. 1995. Importance of GIS to community-based management of wildlife: Lessons from Zambia. *Ecological Applications* 5: 861–872.

Lewis, D. M., G. B. Kaweche and A. Mwenya. 1990. Wildlife conservation outside protected areas—lessons from an experiment in Zambia. *Conservation Biology* 4: 171–180.

Lindberg, K. 1991. *Policies for Maximizing Nature Tourism's Ecological and Economic Benefits*. World Resources Institute, Washington, D.C.

Linden, E. 1994. Ancient creature in a lost world. *Time* (June): 52–54.

Loeschcke, V., J. Tomiuk and S. K. Jain (eds.). 1994. *Conservation Genetics*. Birkhauser Verlag, Basel, Switzerland.

Lovelock, J. 1988. *The Ages of Gaia*. W. W. Norton & Company, New York.

Lubchenco, J., et al. 1991. The sustainable biosphere initiative: An ecological research agenda. *Ecology* 72: 371–412.

Ludwig, D., R. Hilborn and C. Walters. 1993. Uncertainty, resource exploitation and conservation: Lessons from history. *Science* 260: 17, 36.

MacArthur, R. H. and E. O. Wilson. 1967. *The Theory of Island Biogeography*. Princeton University Press, Princeton, NJ.

Mace, G. M. 1994. Classifying threatened species: means and ends. *Phil. Tran. Royal Soc. Lond. B* 344: 91–97.

Mace, G. M. 1995. Classification of threatened species and its role in conservation planning. *In* J. H. Lawton and R. M. May (eds.), *Extinction Rates*, pp. 131–146. Oxford University Press, Oxford.

Mace, G. M. and E. J. Hudson. 1999. Attitudes toward sustainability and extinction. *Conservation Biology* 13: 242–246.

Mace, G. M. and R. Lande. 1991. Assessing extinction threats: Towards a reevaluation of IUCN threatened species categories. *Conservation Biology* 5: 148–157.

MacKenzie, S. H. 1996. *Integrated Resource Planning and Management: The Resource Approach in the Great Lakes Basin*. Island Press, Washington, D.C.

MacKinnon, J. 1983. *Irrigation and watershed protection in Indonesia*. Report to IBRD Regional Office, Jakarta.

Magnuson, J. J. 1990. Long-term ecological research and the invisible present. *BioScience* 40: 495–501.

Makarewicz, J. C. and P. Bertram. 1991. Evidence for the restoration of the Lake Erie ecosystem. *BioScience* 41: 216–223.

Malakoff, D. 1998. Death by suffocation in the Gulf of Mexico. *Conservation Biology* 281: 190–192.

Mangel, M. and C. Tier. 1994. Four facts every conservation biologist should know about persistence. *Ecology* 75: 607–614.

Mann, C. C. and M. L. Plummer. 1993. The high cost of biodiversity. *Science* 260: 1868–1871.

Mann, C. C. and M. L. Plummer. 1999. A species' fate, by the numbers. *Science* 284: 36–37.

Manne, L. L., T. M. Brooks and S. L. Pimm. 1999. Relative risk of extinction of passerine birds on continents and islands. *Nature* 399: 258–261.

Mares, M. A. 1992. Neotropical mammals and the myth of Amazonian biodiversity. *Science* 255: 976–979.

Maser, C. 1997. *Sustainable Community Development: Principles and Practices*. St. Lucie Press, Delray Beach, FL.

Masood, E. and L. Garwin. 1998. Costing the earth: When ecology meets economics. *Nature* 395: 426–430.

Mathews, A. 1992. *Where the Buffalo Roam*. Grove Weidenfeld, New York.

Mathews, S. and M. J. Donoghue. 1999. The root of Angiosperm phylogeny inferred from duplicate phytochrome genes. *Science* 286: 947–950.

May, R. M. 1992. How many species inhabit the Earth? *Scientific American* 267: 42–48.

McCallum, H. and A. Dobson. 1995. Detecting disease and parasite threats to endangered species and ecosystems. *Trends in Ecology and Evolution* 10: 190–194.

McKibben, B. 1996. What good is a forest? *Audubon* 98 (3): 54–65.

McLachlan, J. A. and S. F. Arnold. 1996. Environmental estrogens. *American Scientist* 84: 452–461.

McLean, I. F., A. D. Wight and G. Williams. 1999. The role of legislation in conserving Europe's threatened species. *Conservation Biology* 13: 966–969.

McMichael, A. J., et al. 1999. Globalization and the sustainability of human health. *BioScience* 49: 205–210.

McNeely, J. A., et al. 1990. *Conserving the World's Biological Diversity*. IUCN, World Resources Institute, CI, WWF-US, the World Bank, Gland, Switzerland and Washington, D.C.

McNeely, J. A., J. Harrison, P. Dingwall (eds.). 1994. *Protecting Nature: Regional Reviews of Protected Areas*. IUCN, Cambridge.

Meffe, G. C., C. R. Carroll, and contributors. 1997. *Principles of Conservation Biology*, Second Edition. Sinauer Associates, Sunderland, MA.

Meffe, G. K., A. H. Ehrlich and D. Ehrenfeld. 1993. Human population control: The missing agenda. *Conservation Biology* 7: 1–3.

Menges, E. S. 1990. Population viability analysis for an endangered plant. *Conservation Biology* 4: 52–62.

Menges, E. S. 1991. The application of minimum viable population theory to plants. *In* D. A. Falk and K. E. Holsinger (eds.), *Genetics and Conservation of Rare Plants*, pp. 45–61. Oxford University Press, New York.

Menges, E. S. 1992. Stochastic modeling of extinction in plant populations. *In* P. L. Fiedler and S. K. Jain (eds.), *Conservation Biology: The Theory and Practice of Nature Conservation, Preservation and Management* , pp. 253–275. Chapman and Hall, New York.

Meyer, W. B. and B. L. Turner II. 1994. *Changes in Land Use and Land Cover: A Global Perspective*. Cambridge University Press, New York.

Miller, B., R. P. Reading and S. Forrest. 1996. *Prairie Night: Black-Footed Ferret and the Recovery of Endangered Species*. Smithsonian Institution Press, Washington, D.C.

Miller, G. H., et al. 1999. Pleistocene extinction of *Genyornis newtoni*: Human impact on Australian megafauna. *Science* 283: 205–208.

Miller, K. R. 1996. *Balancing the Scales: Guidelines for Increasing Biodiversity's Chances Through Bioregional Management*. World Resources Institute, Washington, D.C.

Miller, R. M. 1990. Mycorrhizae and succession. *In* W. R. Jordan III, M. E. Gilpin and J. D. Aber (eds.), *Restoration Ecology: A Synthetic Approach to Ecological Research*, pp. 205–220. Cambridge University Press, Cambridge.

Mills, L. S. and F. W. Allendorf. 1996. The one-migrant-per-generation rule in conservation and management. *Conservation Biology* 10: 1509–1518.

Milton, S. J., et al. 1999. A protocol for plant conservation by translocation in threatened lowland Fynbos. *Conservation Biology* 13: 735–743.

Milton, S. J., W. R. J. Dean, M. A. du Plessis and W. R. Siegfried. 1994. A conceptual model of arid rangeland degradation. *BioScience* 44: 70–76.

Minckley, W. L. 1995. Translocation as a tool for conserving imperiled fishes: Experiences in western United States. *Biological Conservation* 72: 297–309.

Mitchell, J. G. 1992. Our disappearing wetlands. *National Geographic* 182 (10): 3–45.

Mitikin, K. and D. Osgood. 1994. *Issues and Options in the Design of Global Environment Facility-Supported Trust Funds for Biodiverisity Conservation*. World Bank, Washington, D.C.

Mittermeier, R. A., N. Myers, P. R. Gil and C. G. Mittermeier. 1999. *Hotspots, Earth's Richest and Most Endangered Terrestrial Ecoregions*. Agrupación Sierra Madre, S.C., Mexico City, Mexico.

Mittermeier, R. A., P. R. Gil and C. G. Mittermeier. 1997. *Megadiversity: Earth's Biologically Wealthiest Nations*. Conservation International, Washington, D.C.

Moffat, M. W. 1994. *The High Frontier: Exploring the Tropical Rainforest Canopy.* Harvard University Press, Cambridge, MA.

Mohsin, A. K. M. and M. A. Ambak. 1983. *Freshwater Fishes of Peninsular Malaysia.* University Pertanian Malaysia Press, Kuala Lumpur, Malaysia.

Moiseenko, T. 1994. Acidification and critical loads for surface waters: Kola, northern Russia. *Ambio* 23: 418–424.

Morell, V. 1994. Serengeti's big cats going to the dogs. *Science* 264: 23.

Morell, V. 1999. The variety of life. *National Geographic* 195 (February): 6–32.

Moyle, P. B. 1995. Conservation of native freshwater fishes in the Mediterranean-type climate of California, USA: A review. *Biological Conservation* 72: 271–279.

Moyle, P. B. and R. A. Leidy. 1992. Loss of biodiversity in aquatic ecosystems: Evidence from fish faunas. *In* P. L. Fiedler and S. K. Jain (eds.), *Conservation Biology: The Theory and Practice of Nature Conservation, Preservation and Management,* pp. 127–169. Chapman and Hall, New York.

Munn, C. A. 1992. Macaw biology and ecotourism or "when a bird in the bush is worth two in the hand". *In* S. R. Beissinger and N. F. R. Snyder (eds.), *New World Parrots in Crisis.* Smithsonian Institution Press, Washington, D.C.

Munn, C. A. 1994. Macaws: Winged rainbows. *National Geographic* 185: 118–140.

Murphy, D. D., D. E. Freas and S. B. Weiss. 1990. An environment-metapopulation approach to population viability analysis for a threatened invertebrate. *Conservation Biology* 4: 41–51.

Murphy, P. G. and A. E. Lugo. 1986. Ecology of tropical dry forest. *Annual Review of Ecology and Systematics* 17: 67–88.

Murrieta, J. R. and R. P. Rueda. 1995. *Extractive Reserves.* 1995 IUCN Forest Conservation Programme. IUCN Publications, Cambridge.

Myers, N. 1986. Tropical deforestation and a mega-extinction spasm. *In* M. E. Soulé, (ed.), *Conservation Biology: The Science of Scarcity and Diversity,* pp. 394–409. Sinauer Associates, Sunderland, MA.

Myers, N. 1987. The extinction spasm impending: Synergisms at work. *Conservation Biology* 1: 14–21.

Myers, N. 1988a. Threatened biotas: "Hotspots" in tropical forests. *Environmentalist* 8: 1–20.

Myers, N. 1988b. Tropical forests: Much more than stocks of wood. *Journal of Tropical Ecology* 4: 209–221.

Myers, N. 1991. The biodiversity challenge: Expanded "hotspots" analysis. *Environmentalist* 10: 243–256.

Myers, N. 1993. Sharing the earth with whales. *In* L. Kaufman and K. Mallory (eds.), *The Last Extinction,* pp. 179–194. MIT Press, Cambridge, MA.

Myers, N. 1996. *Ultimate Security: The Environmental Basis of Political Stability.* Island Press, Washington, D.C.

Myers, N. and N. J. Myers. 1994. *The Primary Source: Tropical Forests and Our Future.* W. W. Norton & Company, New York.

Nabhan, G. P. 1989. *Enduring Seeds: Native American Agriculture and Wild Plant Conservation.* North Point Press, San Francisco.

Naess, A. 1986. Intrinsic value: Will the defenders of nature please rise? *In* M. E. Soulé (ed.), *Conservation Biology: The Science of Scarcity and Diversity,* pp. 153–181. Sinauer Associates, Sunderland, MA.

Naess, A. 1989. *Ecology, Community and Lifestyle.* Cambridge University Press, Cambridge.

Naylor, R. L., et al. 1998. Nature's subsidies to shrimp and salmon farming. *Science* 282: 883–884.

Nei, M., T. Maruyama and R. Chakraborty. 1975. The bottleneck effect and genetic variability in populations. *Evolution* 29: 1–10.

Nepstad, D. C. and S. Schwartzman (eds.). 1992. *Non-Timber Products from Tropical Forests: Evaluation of a Conservation and Development Strategy.* The New York Botanical Garden, Bronx, NY.

Nepstad, D. C., et al. 1999. Large-scale impoverishment of Amazonian forests by logging and fire. *Nature* 398: 505–508.

Neto, R. B. and D. Dickson. 1999. $3 m deal launches major hunt for drug deals in Brazil. *Nature* 400: 302.

Niemelå, J., D. Langor and J. R. Spence. 1993. Effects of clear cut harvesting on boreal ground-beetle assemblages (*Coleoptera: carabidae*) in Western Canada. *Conservation Biology* 7: 551–561.

Norse, E. A. (ed.). 1993. *Global Marine Biological Diversity: A Strategy for Building Conservation into Decision Making*. Island Press, Washington, D.C.

Norton, B. G. 1991. *Toward Unity Among Environmentalists*. Oxford University Press, New York.

Norton, B. G., M. Hutchins, E. F. Stevens and T. L. Maple. 1995. *Ethics on the Ark: Zoos, Animal Welfare, and Wildlife Conservation*. Smithsonian Institution Press, Washington, D.C.

Noss, R. F. 1992. Essay: Issues of scale in conservation biology. *In* P. L. Fiedler and S. K. Jain (eds.). *Conservation Biology: The Theory and Practice of Nature Conservation, Preservation and Management*, pp. 239–250. Chapman and Hall, New York.

Noss, R. F. and A. Y. Cooperrider. 1994. *Saving Nature's Legacy: Protecting and Restoring Biodiversity*. Island Press, Washington, D.C.

Noss, R. F., M. A. O'Connell and D. D. Murphy. 1997. *The Science of Conservation Planning: Habitat Conservation under the Endangered Species Act*. Island Press, Washington, D.C.

Nunney, L. and D. R. Elam. 1994. Estimating the effective population size of conserved populations. *Conservation Biology* 8: 175–184.

O'Brien, S. J. and J. F. Evermann. 1988. Interactive influence of infectious disease and genetic diversity in natural populations. *Trends in Ecology and Evolution* 3: 254–259.

Oates, J. F. 1999. *Myth and Reality in the Rainforest: How Conservation Strategies Are Failing in West Africa*. University of California Press, Berkeley, CA.

Odum, E. P. 1997. *Ecology: A Bridge Between Science and Society*. Sinauer Associates, Sunderland, MA.

Office of Technology Assessment of the U.S. Congress (OTA). 1987. *Technologies to Maintain Biological Diversity*. OTA-F-330. U.S. Government Printing Office, Washington, D.C.

Office of Technology Assessment of the U.S. Congress (OTA). 1993. *Report Brief*. U.S. Government Printing Office, Pittsburgh.

Oldfield, M. L. and Alcorn, J. B. (eds.). 1991. *Biodiversity: Culture, Conservation and Ecodevelopment*. Westview Press, Boulder, CO.

Olney, P. J. S. and P. Ellis (eds.). 1991. *1990 International Zoo Yearbook*, vol. 30. Zoological Society of London, London.

Olson, D. M. and E. Dinerstein. 1998. The Global 200: A representation approach to conserving the Earth's most biologically valuable ecoregions. *Conservation Biology* 12: 502–515.

Olson, S. L. 1989. Extinction on islands: Man as a catastrophe. *In* M. Pearl and D. Western (eds.), *Conservation Biology for the Twenty-first Century*, pp. 50–53. Oxford University Press, Oxford.

Packer, C. 1992. Captives in the wild. *National Geographic* 181 (4): 122–136.

Packer, C. 1997. Viruses of the Serengeti: Patterns of infection and mortality in African lions. *Journal of Animal Ecology* 68: 1161–1178.

Packer, C., A. E. Pusey, H. Rowley, D. A. Gilbert, J. Martenson and S. J. O'Brien. 1991. Case study of a population bottleneck: Lions of the Ngorongoro Crater. *Conservation Biology* 5: 219–230

Palmer, M. E. 1987. A critical look at rare plant monitoring in the United States. *Biological Conservation* 39: 113–127.

Panayotou, T. and P. S. Ashton. 1992. *Not by Timber Alone: Economics and Ecology for Sustaining Tropical Forests*. Island Press, Washington, D.C.

Paoletti, M. G. (ed.). 1999. *Invertebrate Biodiversity as Bioindicators of Sustainable Landscapes*. Elsevier Science, Inc., New York.

Parikh, J. and K. Parikh. 1991. *Consumption Patterns: The Driving Force of Environmental Stress*. UNCED, Geneva, Switzerland.

Parkes, R. J., B. A. Cragg, S. J. Bale, et al. 1994. Deep bacterial biosphere in Pacific Ocean sediments. *Nature* 371: 410–413.

Parmesan, C., et al. 1999. Poleward shifts in geographical ranges of butterfly species associated with regional warming. *Nature* 399: 579–583.

Paton, P. W. C. 1994. The effect of edge on avian nest success: How strong is the evidence? *Conservation Biology* 8: 17–26.

Patterson, A. 1990. Debt for nature swaps and the need for alternatives. *Environment* 32: 5–32.

Pechmann, J. H. K., D. E. Scott, R. D. Semlitsch, et al. 1991. Declining amphibian populations: The problems of separating human impacts from natural fluctuations. *Science* 253: 892–895.

Peres, C. A. and J. W. Terborgh. 1995. Amazonian nature reserves: An analysis of the defensibility status of existing conservation units and design criteria for the future. *Conservation Biology* 9: 34–46.

Perrings, C. 1995. Economic values of biodiversity. *In* V. H. Heywood (ed.), *Global Biodiversity Assessment*, pp. 823–914. Cambridge University Press, Cambridge.

Peterken, G. F. 1996. *Natural Woodland, Ecology and Conservation in Northern Temperate Regions.* Cambridge University Press, Cambridge.

Philippart, J. C. 1995. Is captive breeding an effective solution for the preservation of endemic species? *Biological Conservation* 72: 281–295.

Pimental, D., C. Wilson, C. McCullum, et al. 1997. Economic and environmental benefits of diversity. *BioScience* 47: 747–757.

Pimm, S. L. 1991. *The Balance of Nature?* University of Chicago Press, Chicago.

Pimm, S. L., H. L. Jones and J. Diamond. 1988. On the risk of extinction. *American Naturalist* 132: 757–785.

Pimm, S. L., M. P. Moulton, and L. J. Justice. 1995. Bird extinction in the Central Pacific. *In* J. H. Lawton and R. M. May (eds.), *Extinction Rates*, pp. 75–87. Oxford University Press, Oxford.

Plotkin, M. J. 1993. *Tales of a Shaman's Apprentice.* Viking/Penguin, New York.

Plucknett, D. L., N. J. H. Smith, J. T. Williams and N. M. Anishetty. 1987. *Gene Banks and the World's Food.* Princeton University Press, Princeton, NJ.

Poffenberger, M. (ed.). 1990. *Keepers of the Forest.* Kumarian, West Hartford, CT.

Poiani, K. A., B. D. Richter, M. G. Anderson and H. E. Richter. 2000. Biodiversity conservation at multiple scales: Functional sites, landscapes, and networks. *BioScience* 50: 133–146.

Poole, J. 1996. *Coming of Age with Elephants: A Memoir.* Hyperion, New York.

Popper, F. J. and D. E. Popper. 1991. The reinvention of the American frontier. *Amicus Journal* (Summer): 4–7.

Porter, S. D. and D. A. Savignano. 1990. Invasion of polygyne fire ants decimates native ants and disrupts arthropod communities. *Ecology* 71: 2095–2106.

Poten, C. J. 1991. A shameful harvest: America's illegal wildlife trade. *National Geographic* 180 (9): 106–132.

Power, M. E., D. Tilman, J. A. Estes, B. A. Menge, et al. 1996. Challenges in the quest for keystones. *BioScience* 46: 609–620.

Power, T. M. 1991. Ecosystem preservation and the economy in the Greater Yellowstone area. *Conservation Biology* 5: 395–404.

Prendergast, J. R., R. M. Quinn and J. H. Lawton. 1999. The gap between theory and practice in selecting nature reserves. *Conservation Biology* 13: 484–492.

Prescott-Allen, C. and R. Prescott-Allen. 1986. *The First Resource: Wild Species in the North American Economy.* Yale University Press, New Haven, CT.

Press, D., D. F. Doak and P. Steinberg. 1996. The role of local government in the conservation of rare species. *Conservation Biology* 10: 1538–1548.

Pressey, R. L. 1994. Ad hoc reservations: Forward or backward steps in developing representative reserve systems. *Conservation Biology* 8: 662–668.

Pressey, R. L., C. J. Humphries, C. R. Margules, R. I. Vane-Wright and P. H. Williams. 1993. Beyond opportunism: Key principles for systematic reserve selection. *Trends in Ecology and Evolution* 8:124–128.

Primack, R. B. 1992. Tropical community dynamics and conservation biology. *BioScience* 42: 818–820.

Primack, R. B. 1996. Lessons from ecological theory: Dispersal, establishment and population structure. *In* D. A. Falk, C. I. Millar and M. Olwell (eds.), *Restoring Diversity: Strategies for Reintroduction of Endangered Plants.* Island Press, Washington, D.C.

Primack, R. B. 1998a. *Essentials of Conservation Biology,* Second Edition. Sinauer Associates, Sunderland, MA.

Primack, R. B. 1998b. Monitoring rare plants. *Plant Talk* 15: 29–35.

Primack, R. B. and B. Drayton. 1997. The experimental ecology of reintroduction. *Plant Talk* 11 (October): 25–28.

Primack, R. B. and S. L. Miao. 1992. Dispersal can limit local plant distribution. *Conservation Biology* 6: 513–519.

Primack, R. B. and T. Lovejoy (eds.). 1995. *Ecology, Conservation and Management of Southeast Asian Rainforests*. Yale University Press, New Haven, CT.

Primack, R. B., H. Kobori and S. Mori. In press. Dragonfly pond restoration promotes conservation awareness in Japan. *Conservation Biology*.

Primack, R., D. Bray, H. Galletti and I. Ponciano (eds.). 1998. *Timber, Tourists, and Temples: Conservation and Development in the Maya Forest of Belize, Guatemala, and Mexico*. Island Press, Washington, D.C.

Pritchard, P. C. 1991. "The best idea America ever had": The National Parks service turns 75. *National Geographic* 180: 36–59.

Quammen, D. 1996. *The Song of the Dodo: Island Biogeography in an Age of Extinctions*. Scribner, New York.

Rabinowitz, A. 1993. *Wildlife Field Research and Conservation Training Manual*. International Wildlife Conservation Park, New York.

Radmer, R. J. 1996. Algal diversity and commercial algal products: New and valuable products from diverse algae may soon increase the already large market for algal products. *BioScience* 46: 263–270.

Ralls, K. and J. Ballou. 1983. Extinction: Lessons from zoos. *In* C. M. Schonewald-Cox, S. M. Chambers, B. MacBryde and L. Thomas (eds.), *Genetics and Conservation: A Reference for Managing Wild Animal and Plant Populations*, pp. 164–184. Benjamin/Cummings, Menlo Park, CA.

Ralls, K. and R. L. Brownell. 1989. Protected species: Research permits and the value of basic research. *BioScience* 39: 394–396.

Ralls, K., J. D. Ballou and A. Templeton. 1988. Estimates of lethal equivalents and the cost of inbreeding in mammals. *Conservation Biology* 2: 185–193.

Raup, D. M. 1992. *Extinction: Bad Genes or Bad Luck?* W. W. Norton & Company, New York.

Raven, P. H. and E. O. Wilson. 1992. A fifty-year plan for biodiversity surveys. *Science* 258: 1099–1100.

Reading, R. P. and S. R. Kellert. 1993. Attitudes toward a proposed reintroduction of black-footed ferrets (*Mustela nigripes*). *Conservation Biology* 7: 569–580.

Redfearn, J. 1999. OECD to set up global facility on biodiversity. *Science* 285: 22–23.

Redford, K. H. 1992. The empty forest. *BioScience* 42: 412–422.

Redford, K. H. and B. D. Richter. 1999. Conservation of biodiversity in a world of use. *Conservation Biology* 13: 1246–1256.

Redford, K. H. and C. Padoch (eds.). 1992. *Conservation of Neotropical Rainforests: Working from Traditional Resource Use*. Columbia University Press, Irvington, NY.

Redford, K. H. and J. A. Mansour (eds.). 1996. *Traditional Peoples and Biodiversity Conservation in Large Tropical Landscapes*. The Nature Conservancy, Arlington, VA.

Reed, R. A., J. Johnson-Barnard and W. L. Baker. 1996. Contribution of roads to forest fragmentation in the Rocky Mountains. *Conservation Biology* 10: 1098–1107.

Reid, W. V. 1992. The United States needs a national biodiversity policy. *Issues and Ideas Brief*. World Resources Institute, Washington, D.C.

Reid, W. V. and K. R. Miller. 1989. *Keeping Options Alive: The Scientific Basis for Conserving Biodiversity*. World Resources Institute, Washington, D.C.

Reinartz, J. A. 1995. Planting state-listed endangered and threatened plants. *Conservation Biology* 9: 771–781.

Repetto, R. 1992. Accounting for environmental assets. *Scientific American* 266 (June): 94–100.

Rhoades, R. E. 1991. World's food supply at risk. *National Geographic* 179 (April): 74–105.

Rich, B. 1990. Multilateral development banks and tropical deforestation. *In* S. Head and R. Heinzman (eds.), *Lessons from the Rainforest.* Sierra Club Books, San Francisco.

Rich, B. 2000. Trading in dubious practices: OECD countries must stop export credit agencies funding environmentally damaging and immoral projects. *Financial Times*, February 24, 2000, p. 15.

Richards, S A., H. P. Possingham and J. Tizard. 1999. Optimal fire management for maintaining community diversity. *Ecological Applications* 9: 880–892.

Richman, L. K., et al. 1999. Novel endotheliotropic herpes viruses fatal for Asian and African elephants. *Science* 283: 1171–1176.

Ricketts, T. H., E. Dinerstein, D. M. Olson and C. Loucks. 1999. Who's where in North America? *BioScience* 49: 369–381.

Roberts, D. L., R. J. Cooper and L. J. Petit. 2000. Use of premontane moist forest and shade coffee agroecosystems by army ants in Western Panama. *Conservation Biology* 14: 192–199.

Robinson, J. G., K. H. Redford and E. L. Bennett. 1999. Wildlife harvest in logged tropical forests. *Science* 284: 595–596.

Robinson, M. H. 1992. Global change, the future of biodiversity and the future of zoos. *Biotropica* (Special Issue) 24: 345–352.

Rochelle, J. A., L. A. Lehman and J. Wisniewski (eds.). 1999. *Forest Fragmentation: Wildlife and Management Implications.* Koninkliijke Brill NV, Leiden, Netherlands.

Rogers, D. L. and F. T. Ledig. 1996. *The Status of Temperate North American Forest Genetic Resources.* U.S. Department of Agriculture Forest Service and Genetic Resources Conservation Program, University of California, Davis.

Rolston, H. III. 1989. *Philosophy Gone Wild: Essays on Environmental Ethics.* Prometheus Books, Buffalo, NY.

Rolston, H. III. 1994. *Conserving Natural Value.* Columbia University Press, New York.

Rosenberg, D. K., B. R. Noon and E. C. Meslow. 1997. Biological corridors: Form, function, and efficiency. *BioScience* 47:677–687.

Ruggiero, L. F., G. D. Hayward and J. R. Squires. 1994. Viability analysis in biological evaluations: Concepts of population viability analysis, biological population and ecological scale. *Conservation Biology* 8: 364–368.

Russ, G. R. and A. C. Alcala. 1996. Marine reserves: Rates and patterns of recovery and decline of large predatory fish. *Ecological Applications* 6:947–961.

Sæther, B. 1999. Top dogs maintain diversity. *Nature* 400: 510–511.

Safina, C. 1993. Bluefin tuna in the West Atlantic: Negligent management and the making of an endangered species. *Conservation Biology* 7: 229–234.

Salafsky, N. and R. Margoluis. 1999. Threat reduction assessment: A practical and cost-effective approach to evaluating conservation and development projects. *Conservation Biology* 13: 830–841.

Salonius, P. 1999. Population growth in the United States and Canada: A role for scientists. *Conservation Biology* 13: 1518–1519.

Salwasser, H., C. M. Schonewald-Cox and R. Baker. 1987. The role of interagency cooperation in managing for viable populations. *In* M. E. Soulé (ed.), *Viable Populations for Conservation*, pp. 159–173. Cambridge University Press, Cambridge.

Samways, M. J. 1994. *Insect Conservation Biology.* Chapman and Hall, London.

Sayer, J. A. and S. Stuart. 1988. Biological diversity and tropical forests. *Environmental Conservation* 15: 193–194.

Schaller, G. B. 1993. *The Last Panda.* University of Chicago Press, Chicago.

Schemske, D. W., B. C. Husband, M. H. Ruckelshaus, et al. 1994. Evaluating approaches to the conservation of rare and endangered plants. *Ecology* 75: 584–606.

Schneider, S. 1998. *Laboratory Earth: The Planetary Gamble We Can't Afford to Lose.* Basic Books, New York. Leading authority clearly explains the complex ideas of global climate change and why it is vitally important to take action.

Schonewald-Cox, C. M. 1983. Conclusions: Guidelines to management: A beginning attempt. *In* C. M. Schonewald-Cox, S. M. Chambers, B. MacBryde and L. Thomas (eds.), *Genetics and Conservation: A Reference for Managing Wild Animal and Plant Populations*, pp. 414–445. Benjamin/Cummings, Menlo Park, CA.

Schultes, R. E. and R. F. Raffauf. 1990. *The Healing Forest: Medicinal and Toxic Plants of the Northwest Amazonia*. Dioscorides Press, Portland.

Schulz, H. N., T. Brinkhoff, T. G. Ferdelman, M. H. Marine, A. Teske and B. B. Jorgensen. 1999. Dense populations of a giant sulfur bacterium in Namibian shelf sediments. *Science* 284: 493–495.

Schwartz, M. W. 1999. Choosing the appropriate scale of reserves for conservation. *Annual Review of Ecology and Systematics* 30: 83–108.

Scott, J. M. and B. Csuti. 1996. Gap analysis for biodiversity survey and maintenance. *In* M. L. Reaka-Kudla, D. E. Wilson and E. O. Wilson (eds.), *Biodiversity II: Understanding and Protecting our Biological Resources*, pp. 321–340. John Henry Press, Washington, D.C.

Scott, J. M., B. Csuti and F. Davis. 1991. Gap analysis: An application of Geographic Information Systems for wildlife species. *In* D. J. Decker, M. E. Krasny, G. R. Goff, C. R. Smith and D. W. Gross (eds.), *Challenges in the Conservation of Biological Resources: A Practitioner's Guide*, pp. 167–179. Westview Press, Boulder, CO.

Sea World. 2000. Baleen whale population estimates. *Baleen Whales: Habitat and Distribution, Sea World/Busch Gardens Animal Information Database*. Data from 1996. http://jj.seaworld.org/baleen_whales/estimatesbw.html (April 10, 2000)

Sessions, G. (ed.). 1995. *Deep Ecology for the 21st Century: Readings on the Philosophy and Practice of the New Environmentalism*. Shambala Books, Boston.

Sexton, W. T., R. C. Szaro, N. C. Johnson and A. J. Malik (eds.). 1999. *Ecological Stewardship: A Common Reference for Ecosystem Management*. 3 vols. Elsevier Science Ltd., New York.

Shafer, C. L. 1990. *Nature Reserves: Island Theory and Conservation Practice*. Smithsonian Institution Press, Washington, D.C.

Shafer, C. L. 1995. Values and shortcomings of small reserves. *BioScience* 45: 80–88.

Shafer, C. L. 1997. Terrestrial nature reserve design at the urban/rural interface. *In* M. W. Schwartz (ed.), *Conservation in Highly Fragmented Landscapes*, pp. 345–378. Chapman and Hall, New York.

Shaffer, M. L. 1981. Minimum population sizes for species conservation. *BioScience* 31: 131–134.

Simberloff, D. S. 1992. Do species-area curves predict extinction in fragmented forest? *In* T. C. Whitmore and J. A. Sayer (eds.), *Tropical Deforestation and Species Extinction*. pp. 75–89. Chapman and Hall, London.

Simberloff, D. S. and L. G. Abele. 1982. Refuge design and island biogeographic theory: Effects of fragmentation. *American Naturalist* 120: 41–50.

Simberloff, D. S. and N. Gotelli. 1984. Effects of insularization on plant species richness in the prairie-forest ecotone. *Biological Conservation* 29: 27–46.

Simmons, R. E. 1996. Population declines, variable breeding areas and management options for flamingos in Southern Africa. *Conservation Biology* 10: 504–515.

Skole, D. L., W. H. Chomentowski, W. A. Salas and A. D. Nobre. 1994. Physical and human dimensions of deforestation in Amazonia. *BioScience* 44: 314–322.

Smith, F. D. M., R. M. May, R. Pellew, T. H. Johnson and K. R. Walter. 1993. How much do we know about the current extinction rate? *Trends in Ecology and Evolution* 8: 375–378.

Society for Ecological Restoration. 1991. Program and abstracts, 3rd Annual Conference, Orlando, FL. 18–23 May 1991.

Soltis, P. S. and M. A. Gitzendanner. 1999. Molecular systematics and the conservation of rare species. *Conservation Biology* 13: 471–483.

Soulé, M. (ed.). 1987. *Viable Populations for Conservation*. Cambridge University Press, Cambridge, UK.

Soulé, M. 1985. What is conservation biology? *BioScience* 35: 727–734.

Soulé, M. 1990. The onslaught of alien species and other challenges in the coming decades. *Conservation Biology* 4: 233–239.

Soulé, M. and D. Simberloff. 1986. What do genetics and ecology tell us about the design of nature reserves? *Biological Conservation* 35: 19–40.

Soulé, M. E. and J. Terborgh. 1999. *Continental Conservation: Scientific Foundations of Regional Reserve Networks.* Island Press, Washington, D.C.

Sparrow, H. R., T. D. Sisk, P. R. Ehrlich and D. D. Murphy. 1994. Techniques and guidelines for monitoring neotropical butterflies. *Conservation Biology* 8: 800–809.

Species Survival Commission. 1990. *Membership Directory.* IUCN, Gland, Switzerland.

Spellerberg, I. F. 1994. *Evaluation and Assessment for Conservation: Ecological Guidelines for Determining Priorities for Nature Conservation.* Chapman and Hall, London.

Stanley, T. 1995. Ecosystem management and the arrogance of humanism. *Conservation Biology* 9: 254–262.

Stanley-Price, M. R. 1989. *Animal Re-introductions: The Arabian Oryx in Oman.* Cambridge University Press, Cambridge.

Stattersfield, A. J., M. J. Crosby, A. J. Long and D. C. Wege. 1998. *Endemic Bird Areas of the World: Priorities for Biodiversity Conservation.* BirdLife International, Cambridge.

Stearns, B. P. and S. C. Stearns. 1999. *Watching, From the Edge of Extinction.* Yale University Press, New Haven, CT.

Steer, A. 1996. Ten principles of the New Environmentalism. *Finance and Development* 33 (4): 4–7.

Stein, B. A. and S. R. Flack. 1997. *Species Report Card: The State of U.S. Plants and Animals.* The Nature Conservancy, Arlington, VA.

Stephenson, N. L. 1999. Reference conditions for giant sequoia forest restoration: Structure, process, and precision. *Ecological Applications* 9: 1253–1265.

Stolzenburg, W. 1992. The mussels' message. *Nature Conservancy* 42: 16–23.

Sutherland, W. J. and D. A. Hill. 1995. *Managing Habitats for Conservation.* Cambridge University Press, Cambridge.

Swanson, F. J. and R. E. Sparks. 1990. Long-term ecological research and the invisible place. *BioScience* 40: 502–508.

Szaro, R. C. and D. W. Johnston (eds.). 1996. *Biodiversity in Managed Landscapes: Theory and Practice.* Oxford University Press, New York.

Tangley, L. 1986. Saving tropical forests. *BioScience* 36: 4–15.

Taylor, V. J. and N. Dunstone (eds.). 1996. *The Exploitation of Mammal Populations.* Chapman and Hall, London.

Tear, T. H., J. M. Scott, P. H. Hayward and B. Griffith. 1993. Status and prospects for success of the Endangered Species Act: A look at recovery plans. *Science* 262: 976–977.

Temple, S. A. 1991. Conservation biology: New goals and new partners for managers of biological resources. *In* D. J. Decker et al. (eds.), *Challenges in the Conservation of Biological Resources: A Practitioner's Guide*, pp. 45–54. Westview Press, Boulder, CO.

Templeton, A. R. 1986. Coadaptation and outbreeding depression. *In* M. E. Soulé (ed.), *Conservation Biology: The Science of Scarcity and Diversity*, pp. 105–116. Sinauer Associates, Sunderland, MA.

Terborgh, J. 1974. Preservation of natural diversity: The problem of extinction prone species. *BioScience* 24: 715–722.

Terborgh, J. 1976. Island biogeography and conservation: Strategy and limitations. *Science* 193: 1029–1030.

Terborgh, J. 1999. *Requiem for Nature.* Island Press, Washington, D.C.

Thapa, B. 1998. Debt-for-nature swaps: An overview. *International Journal of Sustainable Development of the World and Ecology* 5: 249–262.

The Nature Conservancy. 1996. *Designing a Geography of Hope: Guidelines for Ecoregion-Based Conservation in The Nature Conservancy.* The Nature Conservancy, Arlington, VA.

Thiollay, J. M. 1992. Influence of selective logging on bird species diversity in a Guianan rain forest. *Conservation Biology* 6: 47–63.

Thomas, C. D. 1990. What do real population dynamics tell us about minimum viable population sizes? *Conservation Biology* 4: 324–327.

Thomas, C. D. and J. C. G. Abery. 1995. Estimating rates of butterfly decline from distribution maps: The effect of scale. *Biological Conservation* 73: 59–65.

Thomas, C. D. and J. Lennon. 1999. Birds extend their ranges northwards. *Nature* 399: 213.

Thornhill, N. W. (ed.). 1993. *The Natural History of Inbreeding and Outbreeding*. University of Chicago Press, Chicago.

Tilman, D. 1999. The ecological consequences of change in biodiversity: A search for general principles. *Ecology* 80: 1455–1474.

Tilman, D., D. Wedin, J. Knops. 1996. Productivity and sustainability influenced by biodiversity in grassland ecosystems. *Nature* 379: 718–720.

Toledo, V. M. 1988. La diversidad biológica de México. *Ciencia y Desarollo*. Conacyt, México City.

Toledo, V. M. 1991. Patzcuaro's lesson: Nature, production and culture in an indigenous region of Mexico. *In* M. L. Oldfield and J. B. Alcorn (eds.), *Biodiversity: Culture, Conservation and Ecodevelopment*, pp. 147–171. Westview Press, Boulder, CO.

Tuljapurkar, S. and H. Caswell (eds.). 1997. *Structured-Population Models and Marine, Terrestrial and Freshwater Systems*. International Thompson Publishing, Florence, KY.

Tunnicliffe, V. 1992. Hydrothermal-vent communities of the deep sea. *American Scientist* 80: 336–349.

Tuxill, J. 1999. *Nature's Cornucopia: Our Stake in Plant Diversity*. World Watch Institute, Washington, D.C.

UNDP/UNEP/World Bank. 1994. *Global Environment Facility: Independent Evaluation of the Pilot Phase*. The World Bank, Washington, D.C.

UNESCO, Division of Ecological Sciences. 1996. *World Network of Biosphere Reserves*. UNESCO, Paris, France.

UNESCO. 1996. *The World Network of Biosphere Reserves*. UNESCO, Paris, France.

Union of Concerned Scientists. 1999. *Global Warming: Early Warning Signs*. Union of Concerned Scientists, Cambridge, MA.

United Nations. 1993a. *Agenda 21: Rio Declaration and Forest Principles*. Post-Rio Edition. United Nations Publications, New York.

United Nations. 1993b. *The Global Partnership for Environment and Development*. United Nations Publications, New York.

United States National Park Service. 2000a. Designation of national park system units. *The National Park Service ParkNet*. Last revised 28 March 2000. http: www.nps.gov/legacy/nomenclature.html (April 12, 2000)

United States National Park Service. 2000b. The national park system acreage. *The National Park Service ParkNet*. Last revised 28 March 2000. http: www.nps.gov/legacy/acreage.html (April 12, 2000)

Valutis, L. L. and J. M. Marzluff. 1999. The appropriateness of puppet-rearing birds for reintroduction. *Conservation Biology* 13: 584–591.

Van de Veer, D. and C. Pierce. 1994. *The Environmental Ethics and Policy Book: Philosophy, Ecology, Economics*. Wadsworth Publishing Company, Belmont, CA.

Van Driesche, R. G. and T. J. Bellows. 1996. *Biological Control*. Chapman and Hall, New York.

Vane-Wright, R. I., C. R. Smith and I. J. Kitching. 1994. A scientific basis for establishing networks of protected areas. *In* P. L. Forey, C. J. Humphries and R. I. Vane-Wright (eds.), *Systematics and Conservation Evaluation*. Oxford University Press, New York.

Veech, J. A. 2000. Choice of species-area function affects identification of hotspots. *Conservation Biology* 14: 140–147.

Vitousek, P. M. 1994. Beyond global warming: ecology and global change. *Ecology* 75: 1861–1876.

Vitousek, P. M., C. M. D'Antonio, L. L. Loope and R. Westerbrooks. 1996. Biological invasions as global environmental change. *American Scientist* 84: 468–478.

Vogel, J. H. 1994. *Genes for Sale: Privatization as a Conservation Policy*. Oxford University Press, New York.

von Droste, B., H. Plachter and M. Rossler (eds.). 1995. *Cultural Landscapes of Universal Value*. Gustav Fischer Verlag, New York.

Wallace, S. S. 1999. Evaluating the effects of three forms of marine reserves on northern abalone populations in British Columbia, Canada. *Conservation Biology* 13: 882–887.

Waller, G. (ed.). 1996. *Sealife: A Guide to Marine Environment*. Smithsonian Institution Press, Washington, D.C.

Wallis deVries, M. F. 1995. Large herbivores and the design of large-scale nature reserves in Western Europe. *Conservation Biology* 9: 25–33.

Ward, D. M., R. Weller and M. M. Bateson. 1990. 16 rRNA sequences reveal numerous uncultured microorganisms in a natural community. *Nature* 345: 63–65.

Ward, G. C. 1992. India's wildlife dilemma. *National Geographic* 181: 2–29.

Watling, L. and E. A. Norse. 1998. Disturbance of the seabed by mobile fishing gear: A comparison to forest clearcutting. *Conservation Biology* 12: 1180–1197.

Weiss, S. B. 1999. Cars, cows, and checkerspot butterflies: Nitrogen deposition and the management of nutrient-poor grasslands for a threatened species. *Conservation Biology* 13: 1476–1486.

Wells, M. and K. Brandon. 1992. *People and Parks: Linking Protected Area Management with Local Communities*. The World Bank/WWF/USAID, Washington, D.C.

Wells, M. P. and K. Brandon. 1993. The principles and practices of buffer zones and local participation in biodiversity conservation. *Ambio* 22: 157–172.

Westemeier, R. L., et al. 1998. Tracking the long-term decline and recovery of an isolated population. *Science* 282: 1695–1698.

Western, D. 1989. Conservation without parks: Wildlife in the rural landscape. *In* D. Western and M. Pearl (eds.), *Conservation for the Twenty-first Century*, pp. 158–165. Oxford University Press, New York.

Western, D. 1997. *In the Dust of Kilimanjaro*. Island Press, Washington, D.C.

Western, D. and J. Ssemakula. 1981. The future of the savannah ecosystem: Ecological islands or faunal enclaves? *African Journal of Ecology* 19: 7–19.

Western, D., R. M. Wright and S. C. Strum (eds.). 1994. *Natural Connections: Perspectives in Community-Based Conservation*. Island Press, Washington, D.C.

White, P. S. 1996. Spatial and biological scales in reintroduction. *In* D. A. Falk, C. I. Millar and M. Olwell (eds.), *Restoring Diversity: Strategies for Reintroduction of Endangered Plants*. Island Press, Washington, D.C.

White, P. S. and J. L. Walker. 1997. Approximating nature's variation: Selecting and using reference information in restoration ecology. *Restoration Ecology* 5: 338–349.

Whitmore, T. C. 1990. *An Introduction to Tropical Rain Forests*. Clarendon Press, Oxford.

Whitten, A. J. 1987. Indonesia's transmigration program and its role in the loss of tropical rain forests. *Conservation Biology* 1: 239–246.

Wijnstekers, W. 1992 *The Evolution of CITES*. CITES Secretariat, Geneva, Switzerland.

Wilcove, D. S. 1999. *The Condor's Shadow: The Loss and Recovery of Wildlife in America*. W. H. Freeman, New York.

Wilcove, D. S. and L. Y. Chen. 1998. Management costs for endangered species. *Conservation Biology* 12: 1405–1407.

Wilcove, D. S. and R. M. May. 1986. National park boundaries and ecological realities. *Nature* 324: 206–207.

Wilcove, D. S., C. H. McLellan and A. P. Dobson. 1986. Habitat fragmentation in the temperate zone. *In* M. E. Soulé (ed.), *Conservation Biology: The Science of Scarcity and Diversity*, pp. 237–256. Sinauer Associates, Sunderland, MA.

Wilcove, D. S., D. Rothstein, J. Dubow, A. Phillips and E. Losos. 1998. Quantifying threats to imperiled species in the United States. *BioScience* 48: 607–615.

Wilcove, D. S., M. McMillan and K. C. Winston. 1993. What exactly is an endangered species? An analysis of the U.S. Endangered Species List: 1985–1991. *Conservation Biology* 7: 87–93.

Wilkes, G. 1991. *In situ* conservation of agricultural systems. *In* M. L. Oldfield and J. B. Alcorn (eds.), *Biodiversity: Culture, Conservation and Ecodevelopment*, pp. 86–101. Westview Press, Boulder, CO.

Wilkinson, C., et al. 1999. Ecological and socioeconomic impacts of 1998 coral mortality in the Indian Ocean: An ENSO impact and a warning of future change? *Ambio* 28: 188–196.

Willers, B. 1994. Sustainable development: A New World deception. *Conservation Biology* 8: 1146–1148.

Williams, J. D. and R. M. Nowak. 1993. Vanishing species in our own backyard: Extinct fish and wildlife of the United States and Canada. *In* L. Kaufman and K. Mallory (eds.), *The Last Extinction*, pp. 107–140. MIT Press, Cambridge, MA.

Wilson, D. E. and R. F. Cole. 1998. *Measuring and Monitoring Biological Diversity: Standard Methods for Mammals*. Smithsonian Institution Press, Washington, D.C.

Wilson, D. E., F. R. Cole, J. D. Nichols and R. Rudran. 1996. *Measuring and Monitoring Biological Diversity: Standard Methods for Mammals*. Biological Diversity Handbook Series. Smithsonian Institution Press, Washington, D.C.

Wilson, D. E., F. R. Cole, J. D. Nichols and R. Rudran. 1996. *Measuring and Monitoring Biological Diversity: Standard Methods for Mammals*. Biological Diversity Handbook Series. Smithsonian Institution Press, Washington, D.C.

Wilson, E. O. 1984. *Biophilia*. Harvard University Press, Cambridge, MA.

Wilson, E. O. 1989. Threats to biodiversity. *Scientific American* 261: 108–116.

Wilson, E. O. 1991. Rain forest canopy: The high frontier. *National Geographic* 180: 78–107.

Wilson, E. O. 1992. *The Diversity of Life*. The Belknap Press of Harvard University Press, Cambridge, MA.

Wilson, E. O. and D. L. Perlmann. 1999. *Conserving Earth's Biodiversity*. Island Press, Washington, D.C.

Wilson, M. A. and S. R. Carpenter. 1999. Economic valuation of freshwater ecosystem services in the United States: 1971–1997. *Ecological Applications* 9: 772–783.

World Conservation Monitoring Centre (WCMC). 1992. *Global Biodiversity: Status of the Earth's Living Resources*. Compiled by the World Conservation Monitoring Centre. Chapman and Hall, London.

World Resources Institute (WRI). 1992. *World Resources 1992–93*. Oxford University Press, New York.

World Resources Institute (WRI). 1994. *World Resources 1994–1995: A Guide to the Global Environment*. Oxford University Press, New York. See also other years.

World Resources Institute (WRI). 1998. *World Resources 1998–99*. Oxford University Press, New York.

Wright, R. G., J. G. MacCracken and J. Hall. 1994. An ecological evaluation of proposed new conservation areas in Idaho: Evaluating proposed Idaho national parks. *Conservation Biology* 8: 207–216

Wright, S. 1931. Evolution in Mendelian populations. *Genetics* 16: 97–159.

Wright, S. J., H. Zeballos, I. Domínguez, M. M. Gallardo, M. C. Moreno and R. Ibañez. 2000. Poachers alter mammal abundance, seed dispersal and seed predation in a Neotropical forest. *Conservation Biology* 14: 227–239.

WWF/IUCN. 1997. *Centres of Plant Diversity: A guide and strategy for their conservation, 3: North America, Middle America, South America, Caribbean Islands*. Cambridge, UK.

Yaffee, S. L. 1999. Three faces of ecosystem management. *Conservation Biology* 13: 713–725.

Yaffee, S. L. et al. 1996. *Ecosystem Management in the United States: An Assessment of Current Experience*. Island Press, Washington, D.C.

Yoakum, J. and W. P. Dasmann. 1971. Habitat manipulation practices. *In* R. H. Giles (ed.), *Wildlife Management Techniques*, pp. 173–231. The Wildlife Society, Washington, D.C.

Young, O. R. (ed.). 1999. *The Effectiveness of International Environmental Regimes: Causal Connections and Behavioral Mechanisms*. MIT Press, Cambridge, MA.

Young, R. A, D. J. P. Swift, T. L. Clarke, G. R. Harvey and P. R. Betzer. 1985. Dispersal pathways for particle-associated pollutants. *Science* 229: 431–435.

Young, T. P. 1994. Natural die-offs of large mammals: Implications for conservation. *Conservation Biology* 8: 410–418.

Zedler, J. B. 1996. Ecological issues in wetland mitigation: An introduction to the forum. *Ecological Applications* 6: 33–37.

Zonneveld, I. S. and R. T. Forman (eds.). 1990. *Changing Landscapes: An Ecological Perspective*. Springer-Verlag, New York.

Index

About the Author

Richard B. Primack is a Professor in the Biology Department at Boston University and is currently a Bullard Fellow at Harvard University (1999–2000). He received his B.A. at Harvard University in 1972 and his Ph.D. at Duke University in 1976. He completed postdoctoral fellowships at the University of Canterbury and Harvard University. His textbook *Essentials of Conservation Biology* (First Edition, 1993; Second Edition, 1998) has also been published in Chinese and German, and *A Primer of Conservation Biology*, First Edition (1995) has been translated, with local coauthors, into Japanese, Indonesian, Vietnamese, and Korean. His other books include *A Field Guide to Poisonous Plants and Mushrooms of North America* (with Charles K. Levy); *A Forester's Guide to the Moraceae of Sarawak*; *Ecology, Conservation and Management of Southeast Asian Rainforests* (edited with Thomas Lovejoy); and *Timber, Tourists, and Temples: Conservation and Development in the Maya Forest of Belize, Guatemala, and Mexico* (edited with David Bray, Hugo Galletti, and Ismael Ponciano).

Dr. Primack's research interests include rare plant conservation and restoration; the ecology, conservation, and management of tropical forests in Southeast Asia and Central America; conservation biology education; and the natural history of orchids. From 1993 to 1999 he was the Book Review Editor for the journal *Conservation Biology*.

About the Book

Editor: Andrew D. Sinauer
Project Editor and Copy Editor: Kerry L. Falvey
Production Manager: Christopher Small
Book Design and Production: Michele Ruschhaupt
Cover Design: Jefferson Johnson
Cover Manufacture: Henry N. Sawyer Company, Inc.
Book Manufacture: Courier Companies, Inc.